Hans-Dieter Junge

**Dictionary of
Artificial Intelligence and
Neuronal Networks**

**Wörterbuch
Künstliche Intelligenz und
neuronale Netzwerke**

© VCH Verlagsgesellschaft mbH, D-6940 Weinheim (Federal Republic of Germany), 1991

Distribution:

VCH, P.O. Box 101161, D-6940 Weinheim (Federal Republic of Germany)

Switzerland: VCH, P.O. Box, CH-4020 Basel (Switzerland)

United Kingdom and Ireland: VCH (UK) Ltd., 8 Wellington Court, Cambridge CB1 1HZ (England)

USA and Canada: VCH, Suite 909, 220 East 23rd Street, New York, NY 10010–4606 (USA)

ISBN 3-527-27994-6 (VCH, Weinheim)
ISBN 0-89573-942-9 (VCH, New York)

ISSN 0930-6862

Hans-Dieter Junge

Dictionary of Artificial Intelligence and Neuronal Networks

English/German
German/English

Wörterbuch Künstliche Intelligenz und neuronale Netzwerke

Englisch/Deutsch
Deutsch/Englisch

Weinheim · New York · Basel · Cambridge

Dr. Hans-Dieter Junge
Cavaillonstr. 78/I
D-6940 Weinheim

This book was carefully produced. Nevertheless, author and publisher do not warrant the information contained therein to be free of errors. Readers are advised to keep in mind that statements, data, illustrations, procedural details or other items may inadvertently be inaccurate.

Published jointly by
VCH Verlagsgesellschaft mbH, Weinheim (Federal Republic of Germany)
VCH Publishers, Inc., New York, NY (USA)

Editorial Director: Dr. Hans-Dieter Junge
Production Manager: Claudia Grössl

Library of Congress Card No. applied for

British Library Cataloguing-in-Publication Data applied for

CIP-Titelaufnahme der Deutschen Bibliothek:
Junge, Hans-Dieter:
Dictionary of artificial intelligence and neuronal networks:
English-German, German-English = Wörterbuch Künstliche
Intelligenz und neuronale Netzwerke / Hans-Dieter Junge. –
Weinheim ; New York ; Basel ; Cambridge : VCH, 1991
  (Parat)
  ISBN 3-527-27994-6 (Weinheim ...)
  ISBN 0-89573-942-9 (New York ...)
NE: HST

© VCH Verlagsgesellschaft mbH, D-6940 Weinheim (Federal Republic of Germany), 1991

Printed on acid-free paper/Gedruckt auf säurefreiem Papier

Alle Rechte, insbesondere die der Übersetzung in andere Sprachen, vorbehalten. Kein Teil dieses Buches darf ohne schriftliche Genehmigung des Verlages in irgendeiner Form – durch Photokopie, Mikroverfilmung oder irgendein anderes Verfahren – reproduziert oder in eine von Maschinen, insbesondere von Datenverarbeitungsmaschinen, verwendbare Sprache übertragen oder übersetzt werden. Die Wiedergabe von Warenbezeichnungen, Handelsnamen oder sonstigen Kennzeichen in diesem Buch berechtigt nicht zu der Annahme, daß diese von jedermann frei benutzt werden dürfen. Vielmehr kann es sich auch dann um eingetragene Warenzeichen oder sonstige gesetzlich geschützte Kennzeichen handeln, wenn sie nicht eigens als solche markiert sind.
All rights reserved (including those of translation into other languages). No part of this book may be reproduced in any form – by photoprinting, microfilm, or any other means – nor transmitted or translated into a machine language without written permission from the publishers. Registered names, trademarks, etc. used in this book, even when not specifically marked as such, are not to be considered unprotected by law.
Composition: U. Hellinger, D-6901 Heiligkreuzsteinach
Printing: Druckerei Winter, D-6900 Heidelberg
Bookbinding: Großbuchbinderei Wilh. Osswald + Co., D-6730 Neustadt/Weinstraße
Printed in the Federal Republic of Germany

Preface

A new age for computers is emerging. Even today most computers still work serially as the old von-Neumann one, and every operation is exactly followed by the next one. The growing field of neuronal networks and neurocomputing represents a radical departure from conventional approaches to digital computers in terms of algorithms and architecture, even though logic is the basic science in both cases. It is highly likely that new software and hardware with powerful computational capabilities will have a significant impact in many fields.

This dictionary gives the translator linguistic equivalents and remarks on their meaning and usage.

Hans-Dieter Junge

Vorwort

Ein neues Zeitalter für den Computer beginnt. Noch arbeiten die meisten Computer seriell, rechnen exakt alle Operationen nacheinander, wodurch immer ein Ablauf exakt auf dem vorangegangenen aufbaut, wie dies vor Jahrzehnten durch von Neumann beschrieben wurde. Die neuronalen Netze bringen jedoch radikale Veränderungen mit sich. Die Algorithmen und die Architektur sind anders, wenn auch weiterhin die Logik Grundlagenwissenschaft bleibt. Es ist mit Sicherheit anzunehmen, daß in den kommenden Jahren neue Software und auch neue Hardware mit mächtigen Möglichkeiten wesentliche Fortschritte machen werden.

Dieses Wörterbuch wurde entwickelt, um dem Übersetzer nicht nur sprachliche Äquivalente zu liefern, sondern auch um ihm zu den wichtigsten Begriffen zusätzliche Hinweise zu ihrer Bedeutung und Verwendung zu geben.

Hans-Dieter Junge

Englisch/Deutsch

A

abbreviation scheme Abkürzungsschema *n* {*Logik, Schaltalgebra*}
Abelian group Abelsche Gruppe *f*, kommutative Gruppe *f*
ability Fähigkeit *f*
abnormal termination [ungewollter] Programmabbruch *m*, Abbruch *m*, Absturz *m*
abort/to abbrechen {*z.B. der Auswertung einer Bitfolge*}, [einen Prozeß] plötzlich beenden
absence of feedback Rückwirkungsfreiheit *f*
absolute absolut
 absolute address absolute Adresse *f*, Absolutadresse *f*, Maschinenadresse *f*
 absolute addressing absolute Adressierung *f*
 absolute assembler absoluter Assembler *m*
 absolute code Maschinencode *m*, Computercode *m*, Rechnercode *m*, absoluter Code *m*
 absolute coding absolute Codierung *f*, Absolutcodierung *f*
 absolute criterion absolutes Kriterium *n*
 absolute extreme value absolutes Extremum *n*, absoluter Extremwert *m*
 absolute optimum absolutes Optimum *n*
 absolute programming Programmieren *n* in Maschinencode {*mit absoluten Adressen*}
 absolute vector absoluter Vektor *m*, Betrag *m* des Vektors
absolutely autonomous system absolut autonomes System *n*,

isoliertes System *n*
abstract 1. abstrakt; 2. Kurzdarstellung *f*, Zusammenfassung *f*, Referat *n* {*Zeitschriftenartikel*}
 abstract automata theory abstrakte Automatentheorie *f*
 abstract automaton abstrakter Automat *m*
 abstract calculus abstrakter Kalkül *m*
 abstract code Pseudocode *m*
 abstract graph abstrakter Graph *m*
 abstract machine abstrakter Automat *m*, abstrakte Maschine *f*
 abstract model abstraktes Modell *n*
 abstract operator abstrakter (begrifflicher) Operator *m*
 abstract space abstrakter (topologischer) Raum *m*
 abstract symbolic model abstraktes (symbolisches) Modell *n*
 abstract syntax abstrakte Syntax *f*
 abstract synthesis of automata abstrakte Synthese *f* von Automaten
abstract/to abstrahieren; suchen; beschreiben, referieren
abstraction Abstraktion *f*
 abstraction level Abstraktionsniveau *n*, Abstraktionsstufe *f*
 abstraction process Abstraktionsprozeß *m*
accelerated gradient method beschleunigte Gradientenmethode *f*
acceleration algorithm Beschleunigungsalgorithmus *m* {*z.B. zur Beschleunigung des Rechenablaufes bei der Gradientenmethode*}
acceptable annehmbar, akzeptabel, akzeptierbar; zulässig
access coding Zugriffscodierung *f*
access destination Zugriffsziel *n*

access mechanism Zugriffsmechanismus
access state Zugriffszustand m
access time Zugriffszeit f
accident zufälliges Ereignis n
accident-sensitive empfindlich gegen zufällige Ereignisse, störanfällig
accidental zufällig; zufallsabhängig
accidental signal zufallsabhängiges Signal n
accomplish/to bewirken, fertigstellen
accretive network Akkretionsnetz n *{neurales Netz zur Gewinnung unverrauschter Daten aus verrauschten}*
accumulate/to akkumulieren, ansammeln
accumulated angesammelt
accumulated error akkumulierter (aufgelaufener) Fehler m
accumulation 1. Akkumulation f, Ansammlung f, Häufung f, Sammlung f, Summierung f, 2. *s.* time-averaging
accumulation point Häufungspunkt m, Grenzpunkt m, Verdichtungspunkt m *{Topologie}*
accumulator 1. Akku m, Akkumulator m, Sekundärelement n; 2. Speicher m *{Teil des Rechenwerks}*; 3. Druckgefäß m; Zähler m; 4. Summenfeld n, Saldierwerk n, *{z.B. als Teil des Rechenwerks}*
accuracy of computation Rechengenauigkeit f
accuracy of resolution Auflösungsgenauigkeit f, Genauigkeit f der Auflösung
accuracy of result Genauigkeit (Richtigkeit) f des Ergebnisses
acknowledge/to bestätigen, quittieren

acknowledgement Bestätigung f, Rückmeldung f, Quittung f, [positive] Antwort f, Richtigbefund m
acoustic dialogue akustischer Dialog m *{Mensch-Maschine-System}*
acoustic input akustische (phonetische) Eingabe f, Spracheingabe f
acoustic output akustische (phonetische) Ausgabe f, Sprachausgabe f,
acquire/to beschaffen, besorgen; bereitstellen, reservieren *{z.B. Speicherplatz}*
act/to bewirken; handeln; verrichten
acting bewirkend; *s.a.* action
action Aktion f; Verhalten n, Wirkungsweise f, Wirkung f; Tätigkeit f, Handlung f, Vorgang m, Einwirkung f, Beeinflussung f
action chain Wirkungskette f
action chart Funktionsdiagramm n
action coalition Handlungskoalition f *{Systemtheorie}*
action field Handlungsfeld n *{Systemtheorie}*
action list Aktionsliste f, Agenda f *{geordnete Liste von Aktionen}*
action part Aktionsteil m *{Befehlssequenz einer Regel, die beim Aufruf das Feuern bewirkt, z.B. bei neuronalen Netzwerken}*
action path Aktionspfad m
action period Abtastzeit f
action phase Abtastzeit f
action potential Aktionspotential n *{Potential zur Charakterisierung der Zustände "erregt" oder "nicht erregt", wenn das Membranpotential den Schwellwert überschritten hat}*
action system Handlungssystem n

action time Vorgangszeit *f*, Zeitdauer *f* eines Vorgangs, Abtastzeit *f* *{Systemtheorie}*
activate/to aktivieren, anregen *{z.B. Schwingungen}*; einschalten, in Gang setzen, auslösen *{z.B. Relais}*
activation Aktivierung *f {Systemtheorie}*; Anregung *f {z.B. von Schwingungen}*, Einschaltung *f*, Auslösung *f {z.B. eines Relais}*
activation button Startknopf *m*
activation function Aktivierungsfunktion *f {gibt an, in welcher Weise das Signal vom Neuron behandelt wird, nachdem der Einfluß der Gewichte vorüber ist}*
activation level Aktivierungsniveau *n*
activation mechanism Aktivierungsmechanismus *m {Mechanismus zum Aufruf einer Prozedur}*
activation state Aktivierungszustand *m {Zustand eines einzelnen Neurons oder eines Neuronennetzes}*
activation value Aktivierungswert *m {Summe der gewichteten Inputs eines Neurons zu einer bestimmten Zeit}*
active aktiv, wirksam, betriebsbereit, funktionsbereit, aktiviert, arbeitend; aufgerufen *{Datei}*; stromführend, spannungsführend; Wirk-
active network aktives Netzwerk *n*
active value aktiver Wert *m {kann während einer Beratung geändert werden}*
activity Aktivität f; Vorgang *m {Netzplantechnik}*
activity analysis Prozeßanalyse *f* mittels Netzplantechnik
activity duration Aktivitätsdauer *f*
activity network Vorgangsnetz *n*
activity number Aktivitätsnummer *f*
activity-on-arrow network vorgangspfeilorientierter Netzplan *m*, Vorgangspfeilnetz *n*
activity-on-node network aktivitätsorientierter Netzplan *m*, Vorgangsknotennetz *n*
activity-oriented network aktivitätsorientierter Netzplan *m*, Vorgangsknotennetz *n*
activity time Aktivitätstermin *m*
actor Wirkungselement *n*
actual wirklich, tatsächlich, effektiv, aktuell, Ist-
actual operating diagram Wirkschema *n*
actual operation tatsächliches Verhalten *n*
actual parameter aktueller Parameter *m*
actual performance tatsächliches Verhalten *n*
actual state Istzustand *m*
actual value Istwert *m {z.B. der Regelgröße}*; aktueller Wert *m*
actuate/to bedienen, betätigen, einschalten, auslösen *{direkt durch den Menschen}*; wirken auf
actuated betätigt
actuating Betätigen *n*, Auslösen *n* *{s.a. actuation}*
actuating element Stellglied *n*, *{s.a. actuator}*
actuating member Betätigungselement *n*
actuating path Wirkungsweg *m*
actuating quantity Betätigungsgröße *f*, Wirkungsgröße *f*

actuating unit Stellglied *n*, Stelleinrichtung *f*, Effektor *m*
actuation Bedienung *f*, Betätigung *f*, Einschalten *n*, Auslösung *f* *{direkt durch den Menschen}*; *s.a.* actuating...
acuity Auflösungsvermögen *n* *{Bild}*;
acyclic *s.* aperiodic
adapt/to adaptieren, anpassen
adaptability Adaptionsfähigkeit *f*, Anpassungsfähigkeit *f* *{Fähigkeit eines Netzwerkes, sich unter sich ändernden Bedingungen selbst zu modifizieren (anzupassen)}*; Verwendbarkeit *f*
adaptable anpassungsfähig, anpaßbar
adaptation Adaptation *f*, Anpassung *f*, Selbstanpassung *f*, Selbsteinstellung *f*, Lernfähigkeit *f*
adaptation algorithm Adaptationsalgorithmus *m*
adaptation law Adaptationsgesetz *n*
adaptation speed Adaptationsgeschwindigkeit *f*
adaption Anpassung *f*, *{s.a.* adaptation*}*
adaptive adaptionsfähig, adaptiv, selbstanpassend, selbsteinstellend, lernfähig
adaptive automaton adaptiver Automat *m*
adaptive channel allocation adaptive Kanalzuweisung *f*
adaptive control system adaptives (selbstregelndes) System *n*, System *n* mit Selbsteinstellung
adaptive control with constraints Grenzwertregelung *f*, Auslastungsregelung *f* *{numerische Steuerung}*
adaptive control with optimization Optimierregelung *f* *{numerische Steuerung}*
adaptive controller adaptiver (selbstanpassender, selbsteinstellender) Regler *m*
adaptive corrector adaptiver Korrektor *m*
adaptive digital element adaptives digitales Element *n*
adaptive filter adaptives Filter *n* *{die Eigenschaften ändern sich automatisch bei sich ändernden Eigenschaften des eintreffenden Signals}*
adaptive learning system adaptives Lernsystem *n*
adaptive line adaptive Fertigungsstraße *f*
adaptive linear network adaptive lineare Schaltung *f*, adaptives Schwellenwertelement *n*
adaptive local optimization algorithm adaptiver lokaler Optimierungsalgorithmus *m*
adaptive logic adaptive (anpassungsfähige, lernfähige) Logik *f*
adaptive loop Adaptionsschleife *f*, Adaptivkreis *m*, anpassungsfähiger Regelkreis *m*
adaptive-loop gain Adaptivkreisverstärkung *f*, Verstärkung *f* des Adaptivkreises
adaptive process line adaptive Fertigungsstraße *f*
adaptive process model adaptives Prozeßmodell *n*
adaptive resonance adaptive Resonanz *f*
adaptive routing adaptive (selbstanpassende) Bahnführung *f*

adaptive sampling adaptive (selbstanpassende) Abtastung *f*
adaptive system adaptives (selbstanpassendes, selbsteinstellendes) System *n*, Adaptivkreis *m*, Adaptionssystem n; *{s.a. adaptive control system}*
adaptive-system modelling Modellierung *f* adaptiver Systeme
adaptive threshold element adaptives Schwellenwertelement *n*
additional logic Zusatzlogik *f*
additional resistance Vorwiderstand *m*, zusätzlicher Widerstand *m* *{physikalische Größe}*
additive additiv
additivity Additivität *f*
addressable adressierbar
addressable horizontal position adressierbare horizontale Position *f*
addressable point adressierbarer Punkt *f*
addressable vertical position adressierbare vertikale Position *f*
adequacy Adäquatheit *f*
adjacency Adjazenz *f*
  adjacency matrix Adjazenzmatrix *f* *{Graphentheorie}*
adjacent adjazent, benachbart; nebeneinander angeordnet
adjoint adjungiert
  adjoint expression adjungierter Ausdruck *m*
  adjoint function adjungierte Funktion *f*
  adjoint matrix adjungierte Matrix *f*
  adjoint operator adjungierter Operator *m*
  adjoint system adjungiertes System *n*
  adjoint system equation adjungierte Systemgleichung *f*, Kozustandsgleichung *f*
  adjoint variable adjungierte Variable *f* *{s.a. costate}*
adjunction Adjunktion *f*
adjustable threshold einstellbarer Schwellenwert *m*
adjusted eingestellt, justiert
adjustment 1. Justierung *f*, Justage *f*, Einstellung *f*, Regulierung *f*, Abgleich *m*; 2. Einstellelement *n* *{s.a. adjustment element}*; 3. Ausgleich *m* *{Ausgleichsrechnung}*; 4. Berichtigung *f*; 5. Aufrundung *f*
  adjustment of conditional observations Ausgleich *m* bedingter Beobachtungen
  adjustment-of-data calculus Ausgleichsrechnung *f*
admissible zulässig
  admissible path zulässiger Weg *m* *{Netzplantechnik}*
  admissible point zulässiger Punkt *m*
  admissible set zulässiges Gebiet *n*; zulässige Menge *f*
  admissible solution zulässige Lösung *f*
  admissible trajectory zulässige Trajektorie *f*
  admissible value zulässiger Wert *m* *{z.B. der Regelabweichung}*
  admissible vector zulässiger Vektor *m*
advance 1. Vorschub *m*, Transport *m*, Weiterbewegung *f*; 2. Voreilung *f* *{Phase}*
  advance angle Voreilwinkel *m*
  advance development Vorentwicklung *f* *{z. B. eines Datensystems}*
advanced modern, fortgeschritten

advanced component modernes Bauelement *n*
advanced logic processing system weiterentwickeltes Logikverarbeitungssystem *n*
advanced operating system weiterentwickeltes Betriebssystem *n*
advanced system modernes System *n*
aeq aeq *{Wahrheitsfunktion GENAU DANN WENN}*
afferent neuron Sensorneuron *n* *{liefert dem Gehirn Signale von anderen Teilen des Körpers}*
after end-point action Signalverlauf *m* nach dem [erwünschten] Endpunkt
age variable Variable *f* der verflossenen Zeit *{Zeitdifferenz zwischen Entstehung und Beobachtung}*
aggregate modelling of large-scale systems Aggregatmodellierung *f* großer Systeme
aggregation Aggregation *f* *{Vereinigung wesentlicher Größen eines Systems}*
AI *s.* artificial intelligence
aim Ziel *n*
aleph hypothesis Alephhypothese *f* *{Verallgemeinerung des Kontinuumproblems}*
algebra Algebra *f*
  algebra of algorithms Algebra *f* der Algorithmen
  algebra of events Ereignisalgebra *f*
  algebra of logic Algebra *f* der Logik
algebraic algebraisch
  algebraic automata theory algebraische Automatentheorie *f*
  algebraic expression algebraischer Ausdruck *m*
  algebraic linguistics mathematische Linguistik *f*
  algebraic notation algebraische Notation *f*, Infix-Notation *f*
  algebraic stability criterion algebraisches Stabilitätkriterium *n*
  algebraic structure algebraische Struktur *f*
  algebraic topology algebraische Topologie *f*
algorithm Algorithmus *m* *{Prozedur zur Lösung von Problemen in einer endlichen Anzahl von Schritten; vorgegebene Regel zur Erreichung eines Zieles}*
  algorithm and program library Algorithmen- und Programmbibliothek *f*
  algorithm scheme Algorithmenschema *n*
  algorithm theory Algorithmentheorie *f*
algorithmic algorithmisch
  algorithmic concept Algorithmenbegriff *m*
  algorithmic logic algorithmische Logik *f*
algorithmization Algorithmierung *f*
  algorithmization of creative processes Algorithmierung *f* kreativer Prozesse
  algorithmization of production processes Algorithmierung *f* von Produktionsprozessen
alias name Andersname *m*, Aliasname *m*
all-or-none law Alles-oder-Nichts-Gesetz *n*
all-powerful method überall anwendbare Methode *f*

allocate/to zuordnen, zuweisen, zuteilen *{z.B. Speicherplätze}*
allocation problem Zuteilungsproblem *n {bei der Optimierung}*
allowable zulässig
almost fast, quasi-
  almost periodic behaviour quasiperiodischer Betriebszustand *m*
ALOA *s.* adaptive local optimization algorithm
alpha language Alpha-Sprache *f*
alpha number Alphanummer *f*, Buchstabennummer *f*
alpha system Alpha-System *n*
alphabet Alphabet *n*, Zeichenvorrat *m*, Repertoire *n*
alphabetic alphabetisch
  alphabetic character Alphabetzeichen *n*, alphabetisches Zeichen *n*
  alphabetic character set alphabetischer Zeichensatz *m*
  alphabetic code alphabetischer Code *m*
  alphabetic operator Alphabetoperator *m*
alternating series alternierende Reihe *f*, Reihe *f* mit abwechselndem Vorzeichen
alternative hypothesis Alternativhypothese *f*
alternator Alternator *m {Systemtheorie}*
ALWAYS IMMER *{Quantor}*
ambiguity Zweideutigkeit *f*, Mehrdeutigkeit *f*, Vieldeutigkeit *f*, Irrelevanz *f {belangloser Teil einer Information}*
ambiguous zweideutig, mehrdeutig, vieldeutig, irrelevant
amount Betrag *m {Mathematik}*; Menge *f*, Umfang *m*, Wert *m*, Größe *f*
amount of computation Rechenaufwand *m*
amount of information Informationsmenge *f*
analog[ue] conclusion Analogieschluß *m*
analog[ue] information Analoginformation *f*, analoge Information *f*
analog[ue] model Analog[ie]-modell *n*
analog[ue] principle Analogieprinzip *n*
analog[ue] relation Analogierelation *f*
analog[ue] simulation Analogsimulierung *f*, analoge Simulierung (Nachbildung) *f*
analogy Analogie *f*
analogy of functions Funktionsanalogie *f*
analysis of automata Analyse *f* von Automaten
analysis of data Datenanalyse *f*, Meßdatenanalyse *f*
analysis of traffic flow Verkehrsanalyse *f*
analysis problem Analysenproblem *n {Netzwerktheorie}*
analysis technique Analyseverfahren *n {z.B. von Systemen}*
analytical continuation analytische Fortsetzung *f*
analytical function analytische Funktion *f*
analytical language model analytisches Sprachmodell *n*
analytical process model analytisches Prozeßmodell *n*

analytical relation analytische Beziehung *f*
analytical solution analytische Lösung *f* *{Systemtheorie}*
analytical transformation analytische Umformung *f*
analyze/to analysieren, auswerten
AND UND
AND circuit UND-Schaltung *f*
AND gate UND-Glied *n*
AND node UND-Knoten *m* *{Knoten im Suchbaum, der erfüllt werden muß}*
AND operation Konjunktion *f*, UND-Verknüpfung *f*
AND/OR graph UND/ODER-Graph *m* *{baumartige Struktur mit UND- und ODER-Knoten}*
animation Animation *f*
ANS *s.* artificial neural system
antagonistic game antagonistisches Spiel *n*
antecedent Antecedens *f* *{linke Seite einer Implikation oder Produktionsregel}*, Vorderglied *n* *{Logik}*; Vorgänger *m*, Vorgeschichte *f*
antinomous set antinomische Menge *f*
antinomy Antinomie *f*, Widerspruch *m* mit sich selbst *{Logik}*
antivalence Antivalenz *f*, ausschließende ODER-Operation *f* *{s.a. exklusive OR}*
antizigzag measure (provision) Antizickzackvorkehrung *f* *{Gradientenmethode}*
apex node Anfangsknoten, Wurzelknoten *m* *{einer Baumstruktur}*
apparatus Apparat *m* *{Bedienungstheorie}*
application Anwendung *f*

applied linguistics angewandte Linguistik *f*
apportioning problem Zuteilungsproblem *n*
approach Methode *f*, Weg *m*, Lösungsweg *m*, Ansatz *m* *{Mathematik}*; Annäherung *f*
approximate annähernd, näherungsweise, etwa, approximativ; *{s.a. approximation}*
approximate determination Näherungsbestimmung *f*
approximate formula Näherungsformel *f*
approximate solution Näherungslösung *f*, angenäherte Lösung *f*
approximate value *s.* approximation value
approximating annähernd, Näherungs-
approximation by iteration Approximation *f* durch Iteration, schrittweise Approximation *f*
approximation calculus Näherungsrechnung *f*
approximation equation Näherungsgleichung *f*
approximation error Approximationsfehler *m*, Fehler *m* durch Approximation
approximation function Näherungsfunktion *f*, Approximationsfunktion *f*
approximation interval Näherungsintervall *n*, Approximationsintervall *n*
approximation method Näherungsmethode *f*, Approximationsmethode *f*
approximation model Näherungsmodell *n*, angenähertes Modell *n*, Approximationsmodell *n*

approximation of function Approximation *f* einer Funktion
approximation polynomial Approximationspolynom *n*, Näherungspolynom *n*
approximation region Näherungsgebiet *n*, Approximationsgebiet *n*
approximation theory Näherungstheorie *f*, Approximationstheorie *f*
approximation value Näherungswert *m*, Approximationswert *m*, Richtwert *m*
approximative näherungsweise
approximative model *s.* approximation model
arbitrary quantity willkürliche Größe *f*
arbitrary-sequence computer logisch fortlaufend organisierter Rechenautomat *m*, Rechenautomat *m* mit beliebiger Befehlsfolge
arbitrary signal willkürliches Signal *n*
arbitration Entscheidung *f*
arbitration logic Entscheidungslogik *f*
arc Bogen *m* *{Verbindung innerhalb eines semantischen Netzes}*
arc evaluation Bogenbewertung *f* *{Netzplanmethode}*
arc of a graph Bogen *m* *{Graph}*
architecture Architektur *f* *{Struktur eines Computersystems oder eines Netzwerks}*
area Fläche *f*, Bereich *m*, Feld *n* *{z.B. eines Speichers}*
arg Argument *n* *{Mathematik}*
argument Argument *n* *{Mathematik}*; Aussagenverbindung *f* *{Logik}*
argument form Argumentation *f* *{Schlußfolgerungsprozedur}*

argument principle Prinzip *n* des Arguments *{z.B. bei Stabilitätskriterien}*
argumentation Argumentation *f* *{logische Schlußprozedur}*
arithmetic arithmetisch, Rechen-
arithmetic element 1. Rechenglied *n*; 2. Rechenelement *n*; 3. arithmetischer Elementarausdruck *m*
arithmetic function arithmetische Funktion *f*, Rechenfunktion *f*
arithmetic operation arithmetische Operation *f*, Rechenoperation *f*
arithmetic operator arithmetischer Operator *m*, Rechenoperator *m*
arithmetic series arithmetische Reihe *f*
arithmetic speed Rechengeschwindigkeit *f*
arithmetical arithmetisch; *{s.a. arithmetic}*
arithmetical and analytical hierarchy arithmetische und analytische Hierarchie *f*
arithmetical logical operations on computer words elementare Wortoperationen *fpl*
arithmetical operations arithmetische Operationen *fpl*
arm positioning Armpositionierung *f* *{Roboter}*
arrangement Anordnung *f*, Aufbau *m*
array Anordnung *f*, Feld *n*, Reihe *f*, Array *n*, Datenfeld *n*, Matrix *f* *{Datenstruktur}*; *s.a.* character string
array declaration Feldvereinbarung *f*
array logic Array-Logik *f*

array of mathematical equations mathematisches Gleichungssystem *n*
array processor Feldprozessor *m*, Arrayprozessor *m*
array store Matrixspeicher *m*
arrival Ankunft *f {Bedienungstheorie}*; Eingang *m*
arrival interval Zwischenankunftszeit *f*
arrival process Ankunftprozeß *m*
arrival rate Ankunftsrate *f*
arrival sequence Ankunftsreihenfolge *f*, Reihenfolge *f* der Ankunft
arrival stream Ankunftstrom *m*, Eingangsstrom *m*, Kundenstrom *m {s.a.* stream of demands}
arrival time Ankunftszeit *f*, Eingangszeit *f*
arriving demand eintreffende Forderung *f {Bedienungstheorie}*
arrow Pfeil *m*
arrow diagram Pfeildiagramm *n {Netzplan}*
arrow-oriented graph pfeilorientierter Netzplan (Graph) *m*
articulated geometry Gelenkgeometrie *f {Roboter}*
artificial künstlich
artificial brain künstliches Gehirn *n*
artificial hand Kunsthand *f*, künstliche Hand *f*
artificial intelligence künstliche Intelligenz, KI, Maschinenintelligenz *f*, Computerintelligenz *f {heuristische Programmierung}*
artificial intelligence programming Programmierung *f* künstlicher Intelligenz
artificial language künstliche Sprache *f*

artificial mapping künstliche Abbildung *f {Umsetzung der Inputmuster in eine innere Darstellung im Netzwerk}*
artificial neural system KI-System *n* mit neuronalem Netzwerk
artificial neuron künstliches Neuron *n*
artificial system künstliches System *n*
ascend/to aufsteigen *{z.B. Reihenfolge}*
ascending order of significance zunehmender Stellenwert *m*
ascent Aufstieg *m*
ascent method Gradientenmethode *f*, Aufstiegsmethode *f*
ascent path Aufstiegsweg *m*, Anstiegsweg *m {Methode des steilsten Aufstiegs}*
ascertainment model Ermittlungsmodell *n*
assembly robot Montageroboter *m*
asserted logisch *{positive Logik}*; logisch 0 *{negative Logik}*
asserted signal aktives Signal *n {H-Signal bei positiver Logik, L-Signal bei negativer Logik}*
assertion 1. Zusicherung *f*, Behauptung *f {Logik}*; 2. aktive Signalebene *f*
assertion logic Aussagenlogik *f*
assign/to zuweisen, zuordnen
assignment Belegung *f*, Zuweisung *f*, Zuordnung *f*
assignment problem Zuordnungsproblem *n*, Ernennungsproblem *n*, Zuteilungsproblem *n*
assistant program Assistenzprogramm *n*
associability Assoziativität *f*

associate/to zuordnen *{Werte}*
association [elementare] Assoziation *f*, Beziehung *f*
associative element assoziatives Element *n*
associative law Assoziativgesetz *n*, assoziatives Gesetz *n*
associative memory network assoziatives Speichernetz *n*
associativity Assoziativität *f*
assumption Voraussetzung *f*, Annahme *f {Logik}*
astable astabil, unstabil, instabil
asymptotic asymptotisch
asymptotic behaviour asymptotisches Verhalten *n*
asymptotic final value asymptotischer Endwert *m*
asymptotic stability asymptotische Stabilität *f {Zustand, bei dem das erwünschte Gleichgewicht erst nach unendlicher Zeit erreicht wird}*
asynchronous automaton asynchroner Automat *m*
asynchronous disconnected mode unabhängiger Wartezustand *m*
asynchronous sequential circuit asynchrones sequentielles Schaltwerk *n*
at-rest signal Null-Signal *n*, logische Null *f*, L-Signal *n {bei positiver Logik}*; H-Signal *n {bei negativer Logik}*; passives Signal *n*
atom Atom *n {Datenelement einer Liste; nicht weiter zerlegbare logische Aussage}*
atomic atomar, Atom-
atomic expression Atomausdruck *m*, atomarer Ausdruck *m*
atomic formula atomare Formel *f*
atomic system atomares System *n*

attack Angriff *m {z.B. auf geheimverschlüsselte Nachricht}*
attenuation Dämpfung *f*, Abschwächung *f*, Schwächung *f*, Abklingen *n {eines Signals}*
attribute Attribut *n {Eigenschaft eines Objektes, z.B. Name oder Alter einer Person}*
attributive grammar Attributgrammatik *f*
audio response unit Sprachausgabe *f*
audio system Tonsystem *n*
authentication Authentifizierung *f*
authentication of messages Authentisierung *f* von Nachrichten
authenticity Authentizität *f*
authentification Authentifikation *f*
autoadaptive selbstanpassend
autoassociative memory network autoassoziatives Speichernetz *n {Eingabe unvollständiger Information führt zur Ausgabe des gesamten Speicherinhaltes}*
autocorrelation Autokorrelation *f*
autocorrelation analysis Autokorrelationsanalyse *f*
autocorrelation function Autokorrelationsfunktion *f*
autocorrelation matrix Autokorrelationsmatrix *f*
autocovariance function Autokovarianzfunktion *f*
autodecrement automatisch zurückrückend
automata language Automatensprache *f*
automata network Netz *n* von Automaten, Automatennetz *n*
automata operator Automatenoperator *m*

automated

automata table Automatentabelle *f*
automata-theoretical microprogram representation automatentheoretische Mikroprogrammdarstellung *f*
automata theory Automatentheorie *f*
automated automatisiert
automated management system automatisiertes Managementsystem *n*
automated system of project design automatisiertes System *n* der Projektierung
automated teaching automatisierter Unterricht *m*
automated theorem proofing (reasoning) automatisiertes Beweisen *n* von Theoremen
automated work organization system automatisiertes System *n* der Arbeitsorganisation
automatic classification automatische Klassifikation *f*
automatic coding automatische Codierung *f*
automatic diagnostics automatische Diagnostik *f*
automatic drafting machine Zeichenautomat *m*
automatic error-correcting system automatisches Fehlerkorrektursystem *n*
automatic error detection automatische Fehlerauffindung *f*
automatic indexing automatische Indizierung *f*, Selbstindizierung *f*
automatic language analysis automatische Sprachanalyse *f*
automatic language translation automatische Übersetzung *f*
automatic reasoning automatisches logisches Schließen *n*; automatische [logische] Beweisführung *f*
automatic speech recognition automatische Spracherkennung *f*
automatic stabilization system System *n* mit automatischer Stabilisierung
automatic syntactical analysis automatische syntaktische Analyse *f*
automatic theorem proofing automatisches Beweisen *n* von Theoremen
automatic translation automatische Übersetzung *f*
automation of management Managementautomatisierung *f*
automation of medical diagnostics Automatisierung *f* der medizinischen Diagnostik
automation of therapy Automatisierung *f* der Therapie
automatization *s.* automation
automaton Automat *m*
automaton without memory Automat *m* ohne Gedächtnis, speicherloser Automat *m*
automorphism Automorphismus *m*
autonomous autonom, selbstständig, unabhängig *{Eigenschaft der selbstständigen Handlungsfähigkeit eines Systems}*
autonomous automaton autonomer Automat *m*
autonomous computer system autonomes Rechnersystem *n*
autonomous system autonomes (unabhängiges, isoliertes) System *n*
auxiliary variable Hilfsvariable *f*
availability Verfügbarkeit *f*
available erhältlich, verfügbar, erreichbar

available signal supply Signalvorrat *m*
available time Zeitvorrat *m*, verfügbare Zeit *f*
average 1. durchschnittlich, gemittelt; 2. Mittelwert *m*, Mittel n; *{s.a. mean}*
average branching factor mittlerer Verzweigungsfaktor *m {Baumstruktur}*
average conditional information content mittlerer bedingter Informationsgehalt *m*, bedingte Entropie (Informationsentropie) *f*
average information content mittlerer Informationsgehalt *m*, Entropie *f*
average information content per symbol mittlerer Informationsbelag *m {bei einer Nachrichtenquelle}*
average information rate per time mittlerer Informationsfluß *m*
average of random function Mittelwert *m* einer Zufallsfunktion
average source entropy Entstehungsgeschwindigkeit *f* von Nachrichten
average transinformation content mittlerer Transinformationsgehalt *m*, Synentropie *f*
average transinformation content per symbol mittlerer Transinformationsbelag *m {einer Nachrichtenquelle}*
average transinformation rate per time mittlerer Transinformationsfluß *m*
average/to Mittelwert bilden, mitteln
axiom Axiom *n*
axiom of choice Auswahlprinzip *n {Mengenlehre}*
axiom of continuity Stetigkeitsaxiom *n*
axiom of extensionality Extensionalitätsaxiom *n*
axiom of infinity Unendlichkeitsaxiom *n*
axiom of regularity Regularitätsaxiom *n*, Fundierungsaxiom *n {Mengenlehre}*
axiom of selection Auswahlaxiom *n*
axiom of the last upper bound Satz *m* von der oberen Grenze
axiom system Axiomensystem *n*
axiomatic axiomatisch
axiomatic calculus of probability axiomatische Wahrscheinlichkeitsrechnung *f*
axiomatic semantics axiomatische Semantik *f*
axon Axon *n {Ansatz an einem biologischen Neuron zum Transport des Ausgangssignals zu anderen Neuronen}*

B

B-tree B-Baum *m*
back propagation Rückwärtsausbreitung *f {Lernmethode, bei der ein durch das Netzwerk zurückgeführtes Fehlersignal die Gewichte verändert, damit der gleiche Fehler nicht noch einmal auftritt}*
back transform Rücktransformierte *f*
back transformation Rücktransformation *f*
back-up system Back-up-System *n*, Reservesystem *n*

backtracking Zurückverfolgen *n*, Rückkehr *f* [zu einem früheren Punkt eines Suchbereiches]; Rückwärtsargumentieren *n* {bei einer zuerst in die Tiefe durchgeführten Suche}
backward chaining Rückwärtsverketten *n* {mit dem Ziel beginnende (deduktive) Suchmethode}; s.a. object-driven
backward reading Rückwärtslesen *n*
bad-defined schlechtdefiniert
balance Gleichgewicht *n*
balance condition Gleichgewichtsbedingung *f*
balance situation Gleichgewichtssituation *f* {Spieltheorie}
balance strategy Gleichgewichtsstrategie *f* {Spieltheorie}
balance/to ins Gleichgewicht bringen; ausgleichen
balanced ausgeglichen, symmetrisch
balanced code gleichgewichtiger Code *m*
balanced state Gleichgewichtszustand *m*
balanced tree ausgeglichener Baum *m*
balk/to umgehen, ausweichen {Bedienungstheorie}
Banach space Banach-Raum *m*
basic concept of decision Grundmodell *n* der Entscheidung
basic connective Grundverknüpfung *f* {Logik}
basic element Grundbaustein *m*, Grundglied *n*, Elementarglied *n*, Grundelement *n*
basic flip-flop Grundflipflop *m*, Basisflipflop *m*, Grund-FF *m*, GFF
basic function Grundfunktion *f*

basic gate Grundgatter *n*
basic interconnection Grundverknüpfung *f* {Logik}
basic logic element logisches Grundelement *n*
basic logic function logische Grundfunktion *f*
basic logic module logischer Grundbaustein *m*
basic matrix Basismatrix *f*
basic set Grundmenge *f* {Mathematik}
basic structure Grundstruktur *f*
basin of attraction Attraktionsbekken *n* {Gebiet um das Energieminimum an der E-Fläche}
Bayes' classifier Bayes-Klassifikator *m* {Spieltheorie}
Bayes' criterion Bayes-Kriterium *n*
Bayes' decision rule Bayes-Entscheidungsregel *f*
Bayes' estimation Bayes-Schätzung *f*
Bayes' teaching Bayessches Lehrverfahren *n*
beginning event Anfangsereignis *n* {Netzplantechnik}
behaving system Verhaltenssystem *n*
behaviour Verhalten *n*, Verlauf *m* {z.B. einer Kurve}, Verhaltensweise *f*
behaviour analysis Verhaltensanalyse *f*
behaviour model Verhaltensmodell *n*
behaviour of automata Verhalten *n* von Automaten
behavioural threshold Verhaltensschwelle *f*
belief Annahme *f*
best case günstigster Fall *m*

bibliographical retrieval bibliographische Suche *f*
bid Gebot *n* {*einer Regel*}
bidirectional associative network bidirectionales assoziatives Netzwerk *n* {*Inputs wirken gleichzeitig als Outputs*}
bifurcate/to sich gabeln
bifurcation Gabelung *f*
bijective function bijektive Funktion *f* {*Mengenlehre*}
bijective homomorphism bijektiver Homomorphismus *m*, Isomorphismus *m*
bimatrix representation Bimatrixdarstellung *f*
bin packing problem Packungsproblem *n*, Behälterpackungsproblem *n*
binary binär, Zweier-, zweistellig {*z.B. Relation*}; zweiwertig, bivalent, dual
binary character Binärzeichen *n*
binary chop *s.* binary search
binary combinational element binäres Verknüpfungsglied *n*
binary decision Binärentscheidung *f*, binäre Entscheidung *f*
binary function *s.* binary logic function
binary information binäre Information *f*
binary logic binäre (zweiwertige) Logik *f*
binary logic function binäre (zweiwertige) logische Funktion *f*, Schaltfunktion *f*, binäre Schaltfunktion *f*
binary matrix Binärmatrix *f*
binary pattern Binärmuster *n*
binary relation binäre Relation *f*
binary search Binärsuche *f*, binäres Suchen *n*
binary source Binärquelle *f*, binäre Informationsquelle *f*
binary structure binäre Struktur *f*
binary system binäres System *n*, Null-Eins-System *n*, Binärsystem *n*, duales Zahlensystem n; {*s.a.* binary number system}
binary tree binärer Baum *m*
binary variable Binärvariable *f*
binomial series expansion Binominalreihenentwicklung *f*, binomische Reihenentwicklung *f*
biological mapping biologische Abbildung *f* {*Gehirnbereiche, die Körperteilen und Sinnesfunktionen entsprechen*}
biological neuron biologisches Neuron *n*
bipolar logic bipolare Logik *f*
birth-and-death process Geburtsund Todesprozeß *m* {*Zufallsprozeß*}
birth process Geburtsprozeß *m* {*Zufallsprozeß*}
bistable 1. bistabil, mit zwei stabilen Zuständen; 2. Multivibrator *m*, Flipflop *m* (*n*)
bistochastic matrix bistochastische Matrix *f*
bit combination Bitkombination *f*
bit configuration Bitmuster *n*, Binärmuster *n*
bit-parallel bitparallel
bit sequence Bitfolge *f*
bit string Bitkette *f*
black-box method Blackbox-Methode *f*, Blackbox-Analyse *f* {*Verfahren zur Untersuchung von nicht bekannten Gliedern*}
blackboard 1. Tafel *f*; 2. Arbeitsspeicher *m* [für die Tafelmethode]

blackboard approach Tafelmethode *f* *{Problemlösungsmethode, bei der die Systemelemente über einen gemeinsamen Arbeitsspeicher miteinander kommunizieren}*
blending problem Mischungsproblem *n* *{Optimierungsproblem}*
blind search blinde Suche *f* *{geordnete Vorgehensweise, die sich nicht auf Wissen stützt}*
block structure Blockstruktur *f*
block world Klötzchenwelt *f* *{eine Kunstwelt aus Blöcken und Pyramiden zur Entwicklung von Vorstellungen z.B. bei der Bilderkennung}*
block/to blockieren
blocking Blockierung *f* *{Sonderfall der Überdeckung}*
Boltzmann machine Boltzmann-Maschine *f* *{ähnlich dem Hopfield-Assoziator, jedoch auf der Grundlage des Vergleiches der Wahrscheinlichkeiten für Netzwerk und Umgebung}*
bond 1. Bond *n* *{Bondgraph}*; 2. Bondstelle *f*
bond graph Bondgraph *m*
bond graph modelling Modellierung *f* mit Bondgraph
Boolean Boolesch, logisch, aussagenlogisch, schaltalgebraisch
Boolean algebra Boolesche Algebra *f*
Boolean combination Boolesche Verknüpfung *f*
Boolean equation logische Gleichung *f*, Gleichung *f* in Boolescher Algebra
Boolean expression Boolescher (logischer) Ausdruck *m*
Boolean function Boolesche (logische) Funktion *f*
Boolean matrix Boolesche Matrix *f*
Boolean operation Boolesche Operation *f*
Boolean operation table Boolesche Operationstafel *f*, Wahrheitswertetafel *f*
Boolean operator Boolescher Operator *m*
Boolean product Boolesches Produkt *n*
Boolean ring Boolescher Ring *m*
Boolean structure Boolesche Struktur *f*, Boolescher Verband *m*
Boolean variable Boolesche Variable *f*, Aussagenvariable *f*, Aussagengröße *f*
Borel field of events Borelsches Ereignisfeld *n*, Borelscher Mengenkörper *m* *{Wahrscheinlichkeitsrechnung}*
bottom-up approach Bottom-up-Methode *f* *{Problemlösung beginnt mit einfachen Aufgaben und schreitet zu komplizierteren Aufgaben fort}*
bottom-up control structure ereignisgesteuerte (datengesteuerte) Steuerstruktur *f* *{Struktur eines Weges zur Problemlösung mit Vorwärtsargumentation aus Anfangsbedingungen}*
bottom-up method Aufbaumethode *f*, Bottom-up-Methode *f* *{Systemtheorie}*
bottom-up strategy Bottom-up-Strategie *f*
bound/to begrenzen *{z.B. auf den stabilen Bereich}*
boundary condition Grenzbedingung *f*, Randbedingung *f*, Randwertbedingung *f*

boundary density function Randdichtefunktion *f {Regression}*
boundary of stability Stabilitätsgrenze *f*
boundary value problem Randwertproblem *n*, Randwertaufgabe *f*, Grenzwertproblem *n*
bounded domain begrenzter Bereich *m*
Box-Wilson searching algorithm Suchalgorithmus *m* von Box-Wilson
brain Gehirn *n*
brain capability Fähigkeit *f* des Gehirnes
brain function Gehirnfunktion *f*
brain network neuronales Netzwerk (Netz) *n*
brainware Brainware *f {Ausbildungsstand, Erfahrung usw. bei der Programmierung}*
branch Abzweig *m*, Zweig *m*, Verzweigung *f {Programm}*; Brückenzweig m; *{s.a. branching}*; Schenkel *m {Thermoelement}*
branch and bound algorithm Verzweigungsalgorithmus *m*
branch and bound method Verzweigungsmethode *f*
branch and bound principle Verzweigungsprinzip *n*
branch method Verzweigungsmethode *f*
branch point Verzweigungspunkt *m*, Verzweigungsstelle *f*, Knotenpunkt *m {Netzwerk}*
branch structure Abzweigstruktur *f*
branch/to sich verzweigen
branching Abzweigung *f*, Verzweigung *f*; *{s.a. branch}*
branching point Verzweigungspunkt *m*, Verzweigungsstelle *f*

branching principle Verzweigungsprinzip *n*
branchpoint *s.* branch point
breadth-first search Breitensuche *f*, Suche *f* in die Breite *{Durchlaufen des Suchbaumes von Ebene zu Ebene}*
break/to unterbrechen *{z.B. ein Programm}*
breakpoint Unterbrechungsstelle *f {z.B. in einem Programm zum Zweck der Fehlersuche}*
browse/to [Text] überfliegen, [Datei] grob durchsehen
building tool Werkzeug *n*
business game ökonomisches Spiel *n*, Spiel *n* in der Planung und Leitung
business graphics Präsentationsgraphik *f*
busy period Besetztperiode *f {Bedienungssystem}*
button Button *m {Zeichen, das nach Anklicken mittels der Maus bestimmte Abläufe bewirkt}*

C

CAI *s.* computer-aided instruction
call Aufruf *m {Datei}*; Call *m {Befehl zur Ausführung einer Funktion, einer einfachen Funktion oder einer Prozedur}*
capacity-restricted transportation problem Transportproblem *n* mit Kapazitätsbeschränkung
cargo-loading problem Ladeproblem *n*, Beladeproblem *n*
Cartesian coordinates kartesische (rechtwinklige) Koordinaten *fpl*

Cartesian robot nach rechtwinkligen Koordinaten arbeitender Roboter *m*
Cartesian space kartesischer Raum *m*
case 1. Fall *m*, Situation *f*; 2. Fallstudie *f*
case study Fallstudie *f*
cause Ursache *f*
centralized zentralisiert, zentral
centre/to zentrieren
certainty Gewißheit *f*, subjektive Sicherheit *f*
certainty factor Konfidenzfaktor *m*
chain 1. Kette *f* {*z.B. zyklische Bitfolge*}; 2. Warteschlange *f*, Schlange *f*
chain/to aneinanderreihen, verketten
chaining Verkettung *f* [von Aussagen]
change Veränderung *f*, Änderung *f*, Wechsel *m*
change in state Zustandsänderung *f*
character Alphazeichen *n*, Zeichen *n*, Symbol *n*; Buchstabe *m*
characteristic 1. Merkmal *n*; 2. Kennziffer *f*
characteristic parameter Kenngröße *f*
characteristic value Kennwert *m*, Merkmalswert *m*, Eigenwert *m*
characteristic values of random variables Kenngrößen *fpl* von Zufallsvariablen
chemical receptor chemischer Rezeptor *m*
chemical transmitter chemischer Transmitter *m*
choice point Knoten *m* des Suchbaumes
chronological chronologisch

chunk Block *m* {*Faktensammlung als Informationseinheit*}
chunk/to zusammenballen {*Stimuli und Verhaltensweisen steuern gemeinsam das Verhalten oder bewirken Ereignisse*}
circuit 1. Schaltung *f*; 2. Übertragungsweg *m*; 3. Umlauf *m*, geschlossener Kreis *m*;
class 1. Klasse {*z.B. Dateiklasse*}; 2. Güteklasse *f*
class assignment Klassenzuweisung *f*, Zuweisung *f* zu einer Klasse {*Objekterkennung*}
class division Klasseneinteilung *f*
class element Klassenelement *n*
class hierarchy Klassenhierarchie *f*
classical conditioning klassische Konditionierung *f* {*eine Lernweise, bei der ein unkonditionierter Stimulus einem konditionierten Stimulus gegenübergestellt wird*}
classify/to klassifizieren
clause Klausel *f* {*eine syntaktische Konstruktion als Teil einer Satzes*}
clear/to 1. freigeben; entlocken; 2. löschen; 3. abschalten {*z. B. bei einem Fehler*}
closed cycle geschlossener Zyklus *m*
cluster Cluster *n* {*einfachste Art von Darstellungen zur Beschreibung von Ideen und Gedanken, z.B. Reihen oder Spalten mit Symbolen*}
CNN *s.* computer neural network
codability Codierbarkeit *f*
cognition Erkennen *n*, Erkenntnis *f*, Kognition *f* {*intellektueller Prozeß*}
cognitive approach Erkenntnismethode *f* {*Lernmethode auf der Basis von Erfahrungen*}

cognitive-philosphical model Erkenntnisphilosophisches Modell *n*
cognitive process Erkenntnisprozeß *m*
cognitive science Erkenntniswissenschaft *f*
coincidence 1. Koinzidenz *f*, Übereinstimmung *f*; 2. [logische] UND-Verknüpfung *f*
column 1. Spalte *f* {*einer Matrix*}; 2. Stelle *f*, Stellenwert *m* {*Zahl*}; 3. Modul *n* {*vertikale Anordnung von Zellen*}
combination Kombination *f*, Verbindung *f*, Verbindungsaufbau m; Verknüpfung *f* {*Logik*}
combinational circuit Verknüpfungsschaltung *f*, Schaltnetz *n*
combinatorial explosion kombinatorische Explosion *f* {*schnelles Anwachsen der Möglichkeiten bei der Vergrößerung eines Suchbereiches*}
combinatorial network Verknüpfungsschaltung *f*
combinatorics Kombinatorik *f*
common sense gesunder Menschenverstand *m* {*Reaktionsfähigkeit des Menschen auf der Basis gesammelter Erfahrungen*}
common-sense reasoning Schlußfolgern *n* [mit dem "gesunden Menschenverstand"], Schließen *n* nach dem gesunden Menschenverstand
commutability Kommutativität *f*
compatibility Kompatibilität *f*, Verträglichkeit *f*
competition Konkurrenz *f*
competitive konkurrierend
compile/to kompilieren {*ein Computerprogramm in eine höhere Sprache übersetzen*}
compiled knowledge kompiliertes (geballtes) Wissen *n*
complete vollständig, komplett
completeness Vollständigkeit *f*
complex komplex, zusammengesetzt; vermascht
complexity 1. Kompliziertheit *f*; 2. Komplexität *f*; Vermaschungsgrad *m*
compliant robot system Roboter *m* mit Kraftkompensation
computability Berechenbarkeit *f*
computational energy function mathematische Energiefunktion *f* {*mathematische Funktion zur Definition stabiler Zustände eines Netzes und der Pfade dahin*}
computational logic Computerlogik *f*
computational tool Rechenhilfe *f*
computational vision maschinelle Bilderkennung *f*, Bilderkennung *f* mit Computer {*wissensgestützter und erwartungsgesteuerter Prozeß zur Interpretation von Modellen aus Sensordaten*}
compute/to [mittels Computer] berechnen
computed berechnet
computed-path control Steuerung *f* nach vorausberechnetem Weg {*Robotersteuerung*}
computer Computer *m*, Rechenautomat *m*, [automatische] Rechenmaschine *f*, [automatischer] Rechner *m*, Rechenanlage *f*, Datenverarbeitungsanlage f; {*s.a. computing...*}
computer-aided instruction computerunterstützte Unterweisung *f*, computerunterstützter Unterricht *m*
computer-aided planning computerunterstützte Planung *f*

computer-aided translation computerunterstützte Übersetzung *f*
computer architecture Computerarchitektur *f* *{Art der Verknüpfung von Rechenelementen zur Realisierung von Computerfunktionen}*
computer-composed music vom Computer komponierte Musik *f*
computer-dependent computerabhängig
computer diagnosis Computerdiagnose *f*, Automatendiagnose *f*
computer generation Computergeneration *f*
computer graphics Computergraphik *f* *{vom Computer generierte visuelle Darstellung}*
computer-independent computerunabhängig
computer language Computersprache *f*
computer linguistics Computerlinguistik *f*
computer network Computernetz *n* *{verbundene Gruppe untereinander kommunizierender Computer}*
computer network architecture Computernetzarchitektur *f*
computer neural network Neurocomputer *m*
computer-readable maschinell lesbar, maschinenlesbar
computer vision maschinelle Bilderkennung *f*, Bilderkennung *f* mit Computer, Bildverständnis *n*, Szenenanalyse *f* *{wissensgestützter und erwartungsgesteuerter Prozeß zur Interpretation von Modellen aus Sensordaten}*
computerized composition computergestützte Zusammensetzung *f*
computerized PC board drilling computergestütztes Bohren *n* von Platinen (Leiterplatten)
computerized question answering computergestützte Fragenbeantwortung *f*
computing automaton Computer *m*, Rechenautomat *m*
computing logic Rechenlogik *f*
computing machine Computer *m*, Rechenautomat *m*; *{s.a. computer}*
computing operation Rechenoperation *f*, arithmetische Operation *f*
computing speed Rechengeschwindigkeit *f*
concatenated data set verkettete Datenmenge *f*
concatenation Verkettung *f* *{von Daten}*
concave function konkave Funktion *f*
conceal/to verbergen, verstecken
concentrate/to 1. konzentrieren *{z.B. Parameter}*; 2. anreichern *{Analysenmeßtechnik}*
concentrated konzentriert
concentrated-parameter network Netzwerk *n* mit konzentrierten Parametern
concept Begriff *m* *{allgemeine Definition}*
concept acquisition Lernen *n* aus Beispielen *{künstliche Intelligenz}*
concept description Konzeptbeschreibung *f* *{symbolische Datenstruktur in der künstlichen Intelligenz}*
concept formation Konzeptbildung *f*, Invariantenbildung *f*
conceptional data structure konzeptionelle Datenstruktur *f*

conceptual begrifflich, konzeptionell
conceptual clustering Anordnen *n* von Objekten nach Klassen gemäß vergegebenen Konzepten
conceptual dependency method Methode *f* der konzeptuellen Abhängigkeit *f* *{Verstehen natürlicher Sprache durch abhängige semantische Elemente}*
conceptual schema konzeptionelles Schema *n*
conclude/to folgern
conclusion Folgerung *f*, Konklusion *f*, Nachsatz *m* *{Logik}*
conclusion-hereditary folgerungserblich
conclusion relation Folgerungsrelation *f*
conclusion rule Transformationsregel *f* *{logische Syntax}*
concordance Konkordanz *f*, Übereinstimmung *f*
concrete automaton konkreter Automat *m*
concurrency logic Gleichzeitigkeitslogik *f*
concurrent nebenläufig, zusammentreffend, gleichzeitig, parallel
concurrent processing überlappte Verarbeitung *f*
concurrent processor Parallelprozessor *m*
concurrent programming Parallelprogrammierung *f*
concurrent working konkurrente (gleichzeitige) Arbeitsweise *f*
condition 1. Bedingung *f*, Kriterium *n*; 2. Zustand *m*, Kondition *f* *{z.B. Grad der Stabilität}*
condition for optimality Optimalitätsbedingung *f*

condition for stability Stabilitätskriterium *n*
condition input Bedingungseingang *m*
condition monitoring Zustandsüberwachung *f*
condition part Bedingungsteil *m* *{logische Bedingung für die Entscheidung, ob eine Regel aktiv ist}*
condition/to vorbereiten, aufbereiten, in einen gewünschten Zustand bringen
conditional bedingt *{soviel wie: NUR DANN, WENN}*
conditional code bedingter Code *m*
conditional connection bedingte Verknüpfung *f*
conditional event bedingtes Ereignis *n*
conditional expectation bedingte Erwartung *f*
conditional expectation value bedingter Erwartungswert *m*
conditional identity rule Satz *m* von der bedingten Identität
conditional implication bedingte Implikation *f*
conditional information content bedingter Informationsgehalt *m*
conditional limited semantics search Suche *f* mit bedingter begrenzter Semantik *{Erkennen von bedingten Mehrwortausdrücken}*
conditional observation bedingte Beobachtung *f*
conditional probability Übergangswahrscheinlichkeit *f*, bedingte Wahrscheinlichkeit *f*
conditional reflex bedingter Reflex *m*

conditional signal bedingtes Signal *n*
conditional stability bedingte Stabilität *f*
conditioned inhibition bedingte Hemmung *f*
conditioned response bedingtes Verhalten (Ansprechen) *n*
conditioned stimulus bedingter Reiz *m*
conditioning konditioniertes Lernen *n*
confidence factor Konfidenzfaktor *m*
configurability Konfigurierbarkeit *f*
configurable konfigurierbar
configurate/to konfigurieren *{Computer}*
configuration Anordnung *f*, Konfiguration *f*, Ausführung[sart] *f*, Konfiguration *f* eines Datenverarbeitungssystems
confirmation Bestätigung *f*
confirmative certainty Sicherheit *f*, Gewißheit *f*
conflict Konflikt *m*
conflict case Konfliktsituation *f*
conflict resolution Konfliktauflösung *f*, Konfliktlösung *f {Strategie zur Wahl einer Instantiation aus der Konfliktmenge}*
conflict set Konfliktmenge *f*
conflict situation Konfliktsituation *f*
conformal konform, übereinstimmend
conformal element konformes Glied *n*
conformal mapping konforme (winkeltreue) Abbildung *f*
conformity Übereinstimmung *f*
congestion Verkehrsstauung *f*; Stauung *f*, Überfüllung *f*; Besetztsein *n*; Verstopfung *f*
congestion theory Bedienungstheorie *f*, Warteschlangentheorie *f*
congruence Kongruenz *f*
congruence relation Kongruenzrelation *f*
congruence system Kongruenzsystem *n*
conjugate gradient konjugierter Gradient *m*
conjugate-gradient method Methode *f* der konjugierten Gradienten
conjugate pole konjugierter Pol *m*
conjunction 1. Konjunktion *f*, logisches Produkt *n*; [logische] Verknüpfung *f*; 2. Konjunktion *f*, logische Multiplikation, UND-Operation *f*
conjunctive normal form konjunktive Normalform *f*
connected automaton zusammenhängender Automat *m*
connected graph zusammenhängender Graph *m*
connected in parallel parallelgeschaltet
connected in series hintereinandergeschaltet, in Serie (Reihe) geschaltet
connectedness of a graph Zusammenhang *m* eines Graphen
connecting line 1. Anschlußleitung *f*; 2. Verknüpfungslinie *f {Graph}*
connecting machine Connecting-Maschine *f*, Parallelcomputer *m*
connecting quantity Verknüpfungsgröße *f {Graph}*
connection 1. Verknüpfung *f*; 2. Zusammenhang *m {z.B. Graph}*
connection matrix Gewichtsmatrix *f*, Entscheidungsmatrix *f* [Matrix

der Connnection-Stärken in einem Hopfield-Netzwerk}, Signalleitung *f* *{entspricht einer Synapse bei einem biologischen Neuron}*
connection point Verknüpfungsstelle *f*
connection strength Bindungsstärke *f*
connection weight Bindungsgewicht *n*
connectionism Konnektionismus *m*
connective instruction Verknüpfungsbefehl *m*
connective operation Verknüpfungsoperation *f* *{Schaltalgebra}*
connective particle Verknüpfungspartikel *n*
connectivity Zusammenhang *m*
connectivity matrix Entscheidungsmatrix *f*; Gewichtsmatrix *f*
consecutive aufeinanderfolgend, fortlaufend
consecutive-sequence computer starr fortlaufend organisierter Computer *m*, Computer *m* mit starrer-Befehlsfolge; *{s.a.* sequential computer*}*
consequence Konsequenz *f* *{Logik}*, logische Folgerung *f*; Folgezustand *m*
consequence symbol Konsequenzzeichen *n*
consequent Folge *f*, Folgerung *f*, [logischer] Schluß *m* *{Resultat der Anwendung eines logischen Operators}*
conservation Erhaltung *f* *{z.B. in der Mathematik}*
conservation law Erhaltungsgesetz *n*, Erhaltungssatz *m*
conservative system konservatives System *n*
conservative value Beharrungswert *m*, Gleichgewichtswert *m*
consistence problem Konsistenzproblem *n*
consistency Konsistenz *f*, Widerspruchsfreiheit *f*
consistency of axiom systems Widerspruchsfreiheit *f* von Axiomensystemen
consistent 1. konsistent *{Schätzung}*; 2. verträglich *{z.B. Randbedingungen}*; 3. zusammen passend, konsistent *{Daten}*
constant 1. Konstante *f*; 2. Symbol *n* mit fester Bedeutung *{Logik}*
constant-coefficient system System *n* mit konstanten Koeffizienten
constant function konstante Funktion *f*
constant-sum condition Konstantsummenbedingung *f*
constant-sum game Konstantsummenspiel *n*
constant with time zeitlich konstant
constituent Bestandteil *m*
constrained begrenzt, eingeschränkt
constrained finite state gebundener Endzustand *m*
constrained maximization problem Maximierungsproblem *n* mit Randbedingungen
constrained miminization problem Minimierungsproblem *n* mit Randbedingungen
constrained optimization Optimierung *f* bei vorgegebenen Randbedingungen
constrained optimization problem Optimierungsproblem *n* mit Randbedingungen

constrained problem s. constrained optimization problem
constrained waiting system Wartesystem *n* mit Beschränkungen
constraint Randbedingung *f*, Zwangsbedingung *f*, Nebenbedingung *f*
constraint-free optimization freie Optimierung *f*, Optimierung *f* ohne Randbedingungen
constraint hypersurface Zwangsbedingungshyperfläche *f*, Hyperfläche *f* der Zwangsbedingungen
constraint of state variables Zustandsbeschränkung *f*
constraint satisfaction Erfüllung *f* von Randbedingungen
constraint search beschränkte Suche *f* {Beschränkung der Suche durch Vorgabe bestimmter Bedingungen}
constraint violation Verletzung *f* von Randbedingungen {Optimierung}
constructional subassembly Baugruppe *f*
constructive analysis konstruktive Analysis *f*
constructive logic konstruktive Logik *f*
consultation Beratung *f*, Konsultation *f*, Abfrage *f* {Expertensystem}
consultation driver Konsultationstreiber *m*
consultation paradigm Beratungsparadigma *n*
contact algebra Kontaktalgebra *f* {ursprüngliche Form der Schaltalgebra}
content Gehalt *m*, Konzentration *f*; Inhalt *m* {z.B. Punktmenge}

content-addressable inhaltsadressierbar
content-analysis Bedeutungsanalyse *f*, Inhaltsanalyse *f*
contention Konkurrenzbetrieb *m*, konkurrierende Anforderung *f* {Informationsübertragung}
context Kontext *m*, Begleittext *m*
context element Kontextelement *n* {Bereich des Arbeitsspeichers, der den Zustand im Programmablauf meldet}
context-free grammar kontextfreie Grammatik *f* {Grammatik zur Beschreibung einer kontextfreien Sprache}
context-free language kontextfreie Sprache *f*
context-sensitive grammar kontextsensitive Grammatik *f*
context tree Kontextbaum *m*, Themenbaum *m*, Objektbaum *m*
contiguous data structure sequentielle Datenstruktur *f*
contingency Kontingenz *f* {Statistik}
continuity Kontinuität *f*, kontinuierlicher Zusammenhang *m*; Stetigkeit *f* {Mathematik}
continuity condition Kontinuitätsbedingung *f*
continuity equation Kontinuitätsgleichung *f*
continuous kontinuierlich, stetig, ununterbrochen, dauernd
continuous decision process stetiger (kontinuierlicher) Entscheidungsprozeß *m*
continuous function stetige Funktion *f*
continuous game kontinuierliches Spiel *n*

continuous graph zusammenhängender Graph *m*
continuous Markovian process stetiger Markow-Prozeß *m*
continuous-path robot streckengesteuerter (bahngesteuerter) Roboter *m*, Stetigbahnroboter *m*
continuous simulation kontinuierliche Simulation *f*
continuous-state neuron stetiges Neuron *n* *{kann jeden beliebigen Wert innerhalb eines Wertebereiches annehmen}*
continuous system stetiges System *n*
continuous-system simulation Simulierung *f* des kontinuierlichen (stetigen) Systems
continuous variable kontinuierliche (analoge, stetige) Größe *f*
continuum problem Kontinuum[s]problem *n*
contraction rule Verschmelzungsregel *f*
contradiction Kontradiktion *f*, [logischer] Widerspruch *m*, Opposition *f*
contradictory kontradiktorisch, widersprüchlich
contradictory form kontradiktorischer Ausdruck *m*
contraposition Kontraposition *f* *{logischer Schluß}*
contrary interest game Spiel *n* mit entgegengesetzten Interessen
contravalence Kontravalenz *f*, Antivalenz *f*, Disjunktion *f*, Exklusiv- ODER n; *{s.a. exclusive OR}*
contravariance Kontravarianz *f*
contravariant kontravariant
control/to 1. steuern; 2. regeln,

regulieren, ausgleichen; 3. lenken, leiten, ansteuern; 4. schalten, betätigen, bedienen, einstellen, verstellen; 5. einhalten *{Sollwert}*; 6. im Griff behalten; geringhalten; einschränken *{z.B. Störungen}*
control Steuerung; Regelung *f*
control hierarchy Steuerungshierarchie *f*, hierarchische Steuerungsstruktur *f*
control method Steuerungsmethode *f* *{z.B. Rückwärtsverkettung oder Vorwärtsverkettung}*
control neuron Steuerneuron *n* *{sendet Signale vom Gehirn zu anderen Teilen des Körpers}*
control objective 1. Steuerungsziel *n*; 2. Regel[ungs]ziel *n*, Regelzweck *m*
control space Steuerraum *m* *{Steuerungstheorie}*
control strategy 1. Steuerstrategie *f*; 2. Regel[ungs]strategie *f*, Schlußfolgerungsstrategie *f*
control structure Steuerstruktur *f* *{Struktur eines Weges zur Problemlösung}*; Schlußfolgerungsstrategie *f* *{Strategie zur Manipulierung von Fachwissen zwecks Lösung eines Problems*
control system Steuer[ungs]system *n*
control system with distributed parameters Steuerungssystem *n* mit verteilten Parametern
control system with variable structure Steuerungssystem *n* mit veränderlicher Struktur
control theory 1. Systemtheorie *f*; 2. Steuerungstheorie *f*; 3. Regelungstheorie *f*

controller logic Ansteuerlogik *f*
convention Vereinbarung *f*
conventional computer konventioneller Computer *m*, von-Neumann-Computer *m*
conventional machine language konventionelle Maschinensprache *f*
conventional machine level konventionelle Maschinenebene *f*
convergence Konvergenz *f*
convergence attribute Konvergenzkriterium *n*
convergence criterion Konvergenzkriterium *n*
convergence proof Konvergenzbeweis *m*
convergence rate Konvergenzgeschwindigkeit *f*
convergent konvergent, konvergierend
convergent domain Konvergenzbereich *m*
conversation Konversation *f*
conversational Dialog-, Interaktiv-
conversational system Dialogsystem *n*
conversational terminal Dialogterminal *n*, Terminal *n* für Dialogbetrieb
conversion 1. Codierung *f*; 2. Umsetzung *f*, Umwandlung *f*, Umcodierung *f* *{Änderung der Darstellung von Daten}*; 3. Umrechnung *f*; 4. Konvertierung *f* *{Übertragung von Daten auf ein anderes Medium}*
convex konvex
convex function konvexe Funktion *f*
convex game konvexes Spiel *n*
convex optimization konvexe Optimierung *f*
convex set konvexe Menge *f*

convolution Faltung *f* *{Mathematik}*
convolution algorithm Faltungsalgorithmus *m*
convolution integral Faltungsintegral *n*
convolution of a function with respect to itself Faltung *f* einer Funktion mit sich selbst
convolution of distributions of random variables Faltung *f* von Verteilungen von Zufallsvariablen
convolution product Faltungsprodukt *n*
convolution sum Faltungssumme *f*
convolution theorem Faltungssatz *m* *{Laplace-Transformation}*
cooperation Kooperation *f*
cooperative game kooperatives Spiel *n*
cooperative game theory kooperative Spieltheorie *f*
cooperative multiperson game kooperatives Mehrpersonenspiel *n*
coordinate/to koordinieren
coordinate system Koordinatensystem *n*, Koordinatenkreuz *n*, Achsenkreuz *n*
coordination Koordination *f*
coordination problem Koordinationsproblem *n*, Zuordnungsproblem *n*, Ernennungsproblem *n*
coordination strategy Koordinierungsstrategie *f* *{Optimierung}*
coordinator Koordinator *m*, Supervisorcomputer *m*
coprocessor Koprozessor *m*
copy/to 1. kopieren, umspeichern, umschreiben; 2. nachahmen *{Lerntheorie}*
correcting capability Korrekturmöglichkeit *f*

corrective korrektiv; Korrektur-
correctness Korrektheit *f* *{z.B. in der Logik}*
corrector Korrektor *m*
correlate/to korrelieren *{Statistik}*
correlated korreliert
correlation Korrelation *f*
correlation computer Korrelationscomputer *m*
correlator Korrelator *m*
correspondence 1. Übereinstimmung *f*; 2. Zuordnung *f*; 3. Korrespondenz *f* *{Mengenlehre}*
corruption Korruption *f* *{Spieltheorie}*
cortex Cortex *m* *{äußerer Teil einer Struktur, z. B. Cerebralcortex}*
counterexample Gegenbeispiel *n*
counterflow Gegenströmung *f*
coupling relation Kopplungsrelation *f*, Kopplungsbeziehung *f*
covariance matrix Kovarianzmatrix *f* *{Wahrscheinlichkeitsrechnung}*
cover/to 1. bedecken, überdecken, erfassen; 2. überstreichen, durchlaufen *{Bereich}*
covering Belegung *f* *{Aussagenvariable}*
covering problem Überdeckungsproblem *n* *{Graphentheorie}*
covering space Überdeckungsraum *m* *{Graph}*
create/to kreieren, erzeugen, generieren
creative schöpferisch, kreativ; künstlerisch
creative computing schöpferische (künstlerische) Computeranwendung *f*
criterion Kriterium *n*
criterion of effectivity Effektivitätskriterium *n*
criterion of judging Entscheidungskriterium *n*, Beurteilungskriterium *n*
criterion of non-interaction Kriterium *n* der Autonomie *{für Beeinflussungsfreiheit}*
criterion of optimality Optimalitätskriterium *n*
criterion of stability Stabilitätskriterium *n*
critical kritisch
critical activity kritische Aktivität *f* *{Netzplan}*
critical event simulation ereignisorientierte Simulation *f*
critical path kritischer Weg *m* *{Netzplan}*
critical path method Methode *f* des kritischen Wegs, CPM-Methode *f*
critical stability kritische Stabilität *f*, Stabilitätsgrenze *f*, Stabilitätsrand *m* *{der Werte in einem Regelungssystem}*
cross section Querschnitt *n*
cue Reiz *m*, Stimulus *m*
current 1. augenblicklich, momentan, laufend; 2. Strom *m*, Stromstärke *f*
current information laufende (momentane) Information *f*
current state laufender (momentaner) Zustand *m*
curve Kennlinie *f*
cut Schnitt *m* *{z.B. Schnittebenenmethode}*
cut-and-try procedure Methode *f* der schrittweisen Näherung
cut-off value Schwellwert *m*
cutpoint network Punktgitternetzwerk *n* *{zellulare Schaltung}*

cutting Schneiden *n*, Zuschneiden *n*
cutting plane Schnittebene *f*
cutting-plane method Schnittebenenmethode *f* *{Integraloptimierung}*
cutting problem Zuschneideproblem *n*, Zuschnittproblem *n*
cybernetic s. cybernetical
cybernetical kybernetisch
cybernetical abstraction kybernetische Abstraktion *f*
cybernetical animal kybernetisches Tier *n*
cybernetical machine kybernetische Maschine *f*
cybernetical model kybernetisches Modell *n*
cybernetical modelling kybernetische Modellierung *f*
cybernetical mouse kybernetische Maus *f*
cybernetical simulator kybernetischer Simulator *m*
cybernetical system kybernetisches System *n*
cybernetical system modelling Modellierung *f* kybernetischer Systeme
cybernetical system theory kybernetische Systemtheorie *f*
cybernetical toy kybernetisches Spielzeug *n*
cybernetical turtle kybernetische Schildkröte *f*
cybernetician Kybernetiker *m*
cybernetics Kybernetik *f*
cycle/to 1. periodisch wiederkehren, sich periodisch wiederholen; 2. schwingen *{z.B. Regelschwingungen}*; 3. kreisen *{Simplexalgorithmus}*

cycle 1. Zyklus *m* *{z.B. Programmzyklus}*; 2. geschlossene Kette *f*, Zyklus *m* *{Graph}*
cycle criterion Schleifenkriterium *n*, Zykluskriterium *n*, Spielzahl *f*, Umlaufzahl *f*
cycle index Schleifenindex *m*, Iterationsindex *m*
cyclic zyklisch, periodisch
cyclic group zyklische Gruppe *f* *{Gruppentheorie}*
cyclic sequence zyklische Folge *f*
cyclic shift zyklische Verschiebung *f*, Ringverschiebung *f*, Registerumlauf *m* *{im Schieberegister}*
cyclical s. cyclic
cycling 1. Regelschwingung *f*; 2. Autorepeat-Funktion *f*, Wiederholfunktion *f*
cyphertext Schlüsseltext *m*, verschlüsselter Text *m*

D

d-chart character Beliebigzeichen *n*
data base Datenbank *f*, Datenbasis *f* *{geordnete Datensammlung zu bestimmtem Sachverhalt}*
data base management system Datenbankverwaltungssystem *n*
data base system Datenbanksystem *n*
data chain Datenkette *f*
data-directed inference datengesteuerte Inferenz *f*
data-driven ereignisgesteuert, datengesteuert *{Problemlösung durch Schlußfolgerungen nach vorwärts und von unten nach oben}*; s.a. forward chaining

data-driven control structure
ereignisgesteuerte (datengesteuerte)
Steuerstruktur *f {Struktur eines Weges zur Problemlösung mit Vorwärtsargumentation aus Anfangsbedingungen}*
data hierarchy Datenhierarchie *f*
data manipulation Datenmanipulation *f*
data manipulation language Datenmanipulationssprache *f*
data range Wertebereich *m*
data structure Datenstruktur *f {Art der Datenspeicherung}*
datum line Bezugslinie *f*
deadlock Systemverklemmung *f {Parallelkonflikt bei von mehreren Prozessoren gemeinsam genutzten Ressourcen}*
death process Todesprozeß *m {Zufallsprozeß}*
debugging Fehlerkorrektur *f {z.B. in einem Programm}*
decentralized dezentralisiert
decentralized intelligence dezentralisierte Intelligenz *f {unterste Ebene von Computerhierarchiesystemen}*
decidability Entscheidbarkeit *f*
  decidability problem Entscheidbarkeitsproblem *n*
decidable entscheidbar
  decidable expression entscheidbarer Ausdruck *m*
decide/to entscheiden
deciding logic Entscheidungslogik *f*
decision Entscheidung *f*
  decision algorithm Entscheidungsalgorithmus *m*
  decision analysis Entscheidungsanalyse *f*

  decision assumption at conditions of indetermination Entscheidungsfindung *f* unter den Bedingungen der Unbestimmtheit
  decision box Blockdiagrammsymbol *n* "Entscheidung", Beschlußkästchen *n {Netzplantechnik}*
  decision circuit Entscheidungsschaltung *f; {s.a.* decision element*}*
  decision content Entscheidungsgehalt *m*, Entscheidungsinhalt *m*
  decision criterion Entscheidungskriterium *n*
  decision device Entscheidungseinheit *f {Gerät}*
  decision domain Entscheidungsbereich *m {Bellmannsche Funktionalgleichung}*
  decision element Entscheidungselement *n*, Entscheidungsglied *n*, Entscheidungsgatter *n*, Bewerter *m*, Entscheidungsschaltung *f*
  decision finding Entscheidungsfindung *f*
  decision function Entscheidungsfunktion *f*
  decision gate *s.* decision element
  decision graph Entscheidungsgraph *m*
  decision instruction Entscheidungsinstruktion *f*, Entscheidungsbefehl *m; {s.a.* branch instruction*}*
  decision integrator Entscheidungsintegrator *m*, inkrementaler Integrierer *m*, Servointegrierer *m*
  decision level Entscheidungsebene *f*
  decision location Entscheidungsstelle *f {z.B. im Petri-Netz}*

decision making Entscheidungsfindung *f*
decision matrix Entscheidungsmatrix *f*
decision method Entscheidungsmethode *f*
decision model Entscheidungsmodell *n*
decision network Entscheidungsnetzplan *m*, Entscheidungsnetz *n*;
{s.a. decision element*}*
decision principle Entscheidungsprinzip *n*
decision probability Entscheidungswahrscheinlichkeit *f*
decision problem Entscheidungsproblem *n*
decision procedure Entscheidungsprozedur *f*, Entscheidungsverfahren *n*
decision process Entscheidungsprozeß *m*
decision program Entscheidungsprogramm *n*
decision rule Entscheidungsregel *f*
decision sequence Entscheidungsablauf *m*, Entscheidungsfolge *f*
decision space Entscheidungsraum *m*
decision stage Entscheidungsstufe *f*
decision strategy Entscheidungsstrategie *f*
decision support system entscheidungsunterstützendes System *n*,
Entscheidungshilfe *f*
decision system Entscheidungssystem *n*, entscheidendes System *n*
decision table Entscheidungstabelle *f*
decision theory Entscheidungstheorie *f*
decision threshold Entscheidungsschwelle *f*
decision tree Entscheidungsbaum *m*
decision tree method Entscheidungsbaumverfahren *n*
decision unit *s.* decision device
decision variable Entscheidungsvariable *f*, Entscheidungsgröße *f*
decision vector Entscheidungsvektor *m*
decision with certainty Entscheidung *f* bei Gewißheit
decision with risk Entscheidung *f*
bei Risiko, Risikoentscheidung *f*
decision with uncertainty Entscheidung *f* bei Ungewißheit
declarative beschreibend; erklärend
**declarative knowledge represen-
tation** erklärende Wissensdarstellung *f* *{Darstellung des Wissens in
Form von Fakten und Behauptungen}*
declarative representation
objektorientierte Darstellung *f*
declarative sentence Aussage *f*
{Logik}; Vereinbarung *f*
decoded text Klartext *m*
decomposability Zerlegbarkeit *f*
decomposability condition Zerlegbarkeitsbedingung *f*
decomposable zerlegbar *{z.B. Aus-
druck}*
**decomposable in propositional
logic** aussagenlogisch zerlegbar
decomposition Dekomposition *f*, Zerlegung *f*; *{s.a.* photodecomposition*}*
decomposition algorithm Dekompositionsalgorithmus *m*
decomposition approach Dekompositionsmethode *f* *{Optimierung}*
decomposition optimization Optimierung *f* durch Dekomposition

decoupling Entkopplung f, Trennung f
dedication Zuordnung f *{einer Ressource auf eine Anwendung}*
deduction Herleitung f, Deduktion, deduktiver Schluß m, Schluß m *{s.a. natural deduction}*
deduction rule Schlußregel f
deduction system Kalkül n, Logikkalkül n
deduction theorem Deduktionstheorem n *{Prädikatenlogik}*
deductive deduktiv, beweisbar, herleitbar
deductive conclusion deduktive Folgerung f
deductive fault analysis deduktive Fehleranalyse f, Fehlerbaumanalyse f
deductive inference deduktiver Schluß m
deductive shell deduktives Expertensystem n
deductive system deduktives System n, Deduktionssystem n
deep knowledge Tiefensuche f
default value aktiver Wert m; Standardwert m *{vordefinierter Wert für den Fall, daß nicht ausdrücklich ein anderer Wert zu verwenden ist}*, Vorgabewert m, vorgegebener Wert m *{dann zu verwendender Wert, wenn der tatsächliche Wert nicht bekannt ist}*
definable definierbar
define/to bestimmen, definieren
defined definiert
defining equation Bestimmungsgleichung f
definite definit, abgegrenzt, bestimmt, vorgegeben
definite automaton definiter Automat m
definite event definiertes Ereignis n
definite game bestimmtes Spiel n
definition Definition f
definition by cases Fallunterscheidung f
degenerate/to entarten
degenerate game entartetes Spiel n
degenerated system entartetes System n
degeneration 1. Degeneration f, Entartung f *{z.B. lineare Optimierung}*; 2. *s.* degenerative feedback
degree Grad m *(n)*
degree of a node Ordnung f eines Knotens
degree of a tree Ordnung f eines Baums
degree of abstraction Abstraktionsgrad m
delay Verzögerung f
delta rule Deltaregel f *{eine Lernregel}*
demand 1. Bedarf m *{z.B. in der Lagerhaltungstheorie}*; 2. Forderung f *{Bedienungstheorie}*
demand predicition Bedarfsvorhersage f *{Lagerhaltungstheorie}*
demand priority Forderungspriorität f, Dringlichkeit f der Forderung *{Bedienungstheorie}*
demand stream Forderungsstrom m *{Bedienungstheorie}*
demented program sinnloses Programm n
demon Dämon m
demonstration Beweisführung f
denary logic zehnwertige (denäre) Logik f
dendrite Dendrit n *{Eingangskanal des Neurons}*

density 1. Dichte *f*; 2. Schwärzung *f* {*Film*}
density function Verteilungsdichtefunktion *f*, Dichtefunktion *f*
denumerable set abzählbare Menge *f*
dependence Abhängigkeit *f*
dependence graph Abhängigkeitsgraph *m*
dependency Abhängigkeit *f*
dependency-directed abhängigkeitsgesteuert
dependency-directed backtracking relevante Rückwärtsverkettung *f*
depression Depression *f*
depth Tiefe *f*
depth cue Tiefeninformation *f*
depth-first search Tiefensuche *f*, Suche *f* in die Tiefe {*Suche von Knoten zu Knoten bis zur Lösung oder bis ein bereits dagewesener Knoten wieder erreicht wird*}
deque *s*. double-ended queue [programm]
dequeue/to aus der Warteschlange entfernen
derive by reasoning/to [logisch] schließen
descent method Abstiegsmethode *f*
descent path Abstiegsweg *m* {*Abstiegsmethode*}
describability Beschreibbarkeit *f*
describable beschreibbar
describe/to beschreiben
description Beschreibung *f*; Objektbeschreibung *f*
description language Beschreibungssprache *f*
description level Beschreibungsebene *f*
description of the recognition object Beschreibung *f* des Erkennungsobjekts
descriptive language deskriptive (beschreibende) Sprache *f*
descriptive logic deskriptive Logik *f*
descriptive model beschreibendes Modell *n*
descriptor Deskriptor *m*
descriptor definition language Deskriptor-Definitionssprache *f*
design/to entwerfen, konstruieren
design 1. Plan *m*, Entwurf *m*; Konzept *n*; 2. Planung *f*, Versuchsplanung *f*; 3. Struktur *f*
design algorithm Entwurfsalgorithmus *m*
design automation Entwurfsautomatisierung *f*
design strategy Entwurfsstrategie *f*
design theory Entwurfstheorie *f*
desired law vorgegebene Gesetzmäßigkeit *f*
destination Ziel *n*, Zielort *m*, Empfangsort *m*, Bestimmungsort *m*
destination data Zieldaten *npl*
detectability Erkennbarkeit *f*
detention time 1. Verweilzeit *f*, Rückhaltzeit *f*; 2. Bedienungsdauer *f* {*Bedienungstheorie*}
determinacy Determiniertheit *f*
determinated automaton determinierter Automat *m*
determinated machine Turing-Maschine *f*, Turing-Automat *m*
determinated system determiniertes System *n*, Turing-System *n*
determination Bestimmung *f*, Feststellung *f*, Ermittlung *f*
deterministic determiniert, deterministisch, kausal bedingt

deterministic algorithm deterministischer Algorithmus *m*
deterministic automaton deterministischer (determinierter) Automat *m*
deterministic chaos determiniertes Chaos *n*, determinierte Unordnung *f*
deterministic disorder determinierte Unordnung *n*, determiniertes Chaos *n*
deterministic optimization deterministische Optimierung *f*
deterministic optimizing problem deterministisches Optimierungsproblem *n*
deterministic process determinierter Prozeß *m*
deterministic signal deterministisches Signal *n*
deterministic simulation deterministische Simulation *f*
deterministic system deterministisches (determiniertes) System *n*
development tool Entwicklungswerkzeug *n* {z.B. für Software}
developmental model Entwicklungsmodell *n*, Brettschaltung *f*, Versuchsmodell *n*
diagnosable diagnostizierbar
diagnosis Diagnose *f*
diagnosis computer Diagnosecomputer *m*, Diagnoseautomat *m*
diagnostic computer Diagnoseautomat *m*, Diagnosemaschine *f*
diagnostic paradigm Diagnoseparadigma *n*
diagnostic program Diagnoseprogramm *n*
diagonal matrix Diagonalmatrix *f*
dialect Dialekt *m* {z.B. einer Computersprache}
dialog[ue] Dialog *m*, Informationsaustausch *m*
dialog[ue] language Dialogsprache *f*
dialog[ue] system Dialogsystem *n*
dialog[ue] unit Dialoggerät *n*
dichotomic classifier dichotomer Klassifikator *m*
dichotomic sequential search dichotome sequentielle Suche *f*
dichotomizing search binäres Suchen *n*, Einstichverfahren *n*, eliminierende Suche *f*, dichotomische Suche *f*, bisektionelles Suchen *n*
dichotomy Dichotomie *f*
dictionary 1. Wörterbuch *n*; 2. Dateiliste *f*, Dateiverzeichnis *n* {Computer}
dictionary retrieval Wörterbuchsuche *f*
diet problem Diätproblem *n*, Problem *n* der minimalen Futterkosten
difference Differenz *f*
difference reduction Differenzverringerung *f*, Reduzierung *f* des [restlichen] Unterschiedes {Problemlösungsmethode, mittels wiederholter Anwendung von Operatoren zu einer Lösung zu gelangen, wobei die Abweichung zwischen aktuellem Zustand und Ziel immer mehr abnimmt}
dimension Dimension *f* {Anzahl der Komponenten eines Vektors}
direct optimization direkte Optimierung *f*, Optimierung *f* ohne Modell
directed gerichtet
directed edge gerichtete Kante *f*, Bogen *m* {Graph}
directed graph gerichteter

discernible 36

Graph *m*, Digraph *m* *{Graph zur Darstellung eines Wissensstruktur, der zwischen den Knoten gerichtete Verbindungslinien hat}*
discernible event unterscheidbares Ereignis *n*
discrete diskret
discrete automaton diskreter Automat *m*
discrete deterministic multistage decision process diskreter deterministischer Mehrstufenentscheidungsprozeß *m*
discrete maximum principle diskretes Maximumprinzip *n*
discrete optimization (optimizing) diskrete (ganzzahlige) Optimierung *f*
discrete-time model Diskretzeitmodell *n*, zeitdiskretes Modell *n*
discrimination Unterscheidung *f* in der Verhaltensweise *{wenn bei zwei vorhandenen Stimuli einer im Vergleich zum anderen stärker ist und damit ein bestimmtes Verhalten bewirkt}*
disjoint disjunkt, elementfremd; durchschnittsfremd *{Mengenlehre}*
disjoint set disjunkte Menge *f*
disjunction Disjunktion *f*, logische Summe *f*, einschließendes ODER *n*, NOR-Operation *f*; *{s.a. exclusive OR}*
disjunctive disjunktiv
display/to [an]zeigen, darstellen
dispose/to Speicherplatz wieder freigeben *{z.B. bei dynamischen Objekten}*
disprove/to widerlegen *{Hypothese}*
disproving Widerlegung *f*, Unmöglichkeitsbeweis *m* *{Hypothese}*
distinguish/to unterscheiden

distributed verteilt
distributed computer network Computernetzwerk *n* mit verteilter Intelligenz
distributed control verteilte Steuerung *f*
distributed data processing Datenverarbeitung *f* mit verteilter Intelligenz
distributed intelligence verteilte Intelligenz *f* *{Computernetzwerk}*
distributed-intelligence computer system Mehrcomputersystem *n*, Computersystem *n* mit verteilter Intelligenz
distributed-intelligence microcomputer system Mikrocomputersystem *n* mit verteilter Intelligenz
distributed-intelligence system System *n* mit verteilter Intelligenz
distributed logic verteilte Logik *f*
distributed memory verteiltes Gedächtnis *n* *{nicht an einer einzelnen Adresse sondern über das gesamte Parallsystem verteilt}*
distributed system verteiltes System *n*, System *n* mit verteilter Intelligenz
distribution law 1. Verteilungsgesetz *n*; 2. distributives Gesetz *n*
distribution problem Distributionsproblem *n*, Verteilungsproblem *n*, Transportproblem *n*
distributive law Distributivgesetz *n*, distributives Gesetz *n*
distributivity Distributivität *f*
disturbed gestört
disturbed information gestörte Information *f*
disturbed state gestörter Zustand *m*

disturbed system gestörtes System *n*
disturbing Stör-
disturbing effect Störeffekt *m*
disturbing noise störendes Rauschen *n*
dit Dit *n {Kurzzeichen dit, Einheit für den Informationsgehalt H bei N verschiedenen Symbolen, definiert durch $H = 10 \log N$}*
divergent divergent, divergierend
divergent series divergierende Reihe *f*
domain Bereich *m*; [interessierender] Problembereich *m*; Wissensgebiet *n*; Definitionsbereich *m*, Gültigkeitsbereich *m*; Domäne *f {Feld im Datensatz, Themenbereich oder Wissensgebiet}*, Interessenbereich *m*
domain expert Experte *m* auf Spezialgebiet, Spezialist *m*
domain of influence Wirkungsbereich *m {z.B. eines Quantors}*
domain of interpretation Individuenbereich *m*
double-ended queue [program] Warteschlangenprogramm *n* für die Bearbeitung einer Liste von beiden Enden her
drive Trieb *m {biologisch wesentlicher Reiz, z.B. Hunger, Angst}*
driving by events *s.* event-driven
DSS *s.* decision support system
dual 1. dual; widerstandsreziprok *{z.B. Zweipol}*; 2. doppelt; *{s.a. double}*; 3. zweiwertig
dual Boolean function duale Boolesche Funktion *f*
dual gradient method Methode *f* der doppelten Gradienten
dual graph Doppelgraph *m*

dual operation duale Operation *f*
dual problem Dualproblem *n*, duales Problem *n {Simplexmethode}*
dual problem of convex optimization duale Aufgabe *f* der konvexen Optimierung
dual simplex algorithm dualer Simplexalgorithmus *m*
duality Dualität *f*
duality principle Dualitätsprinzip *n*
duel Duell *n {Spieltheorie}*
duel condition Duellsituation *f {Spieltheorie}*
dumb terminal dummes Terminal *n {als Gegensatz zum intelligenten Terminal}*
duplicator Vervielfältiger *m*, Kopierer *m*; Kopiator *m {Systemtheorie}*
dyadic Boolean operation dyadische Boolesche Operation *f*
dynamic dynamisch, sich zeitlich ändernd
dynamic game dynamisches Spiel *n*
dynamic knowledge base dynamische Wissensbank *f*, Kurzzeitgedächtnis *n*
dynamic model dynamisches Modell *n*
dynamic-priority queueing system Bedienungssystem *n* mit dynamischer Priorität
dynamic process model dynamisches Prozeßmodell *n*
dynamic stability dynamische Stabilität *f {Fähigkeit eines neuronalen Netzwerkes, nach einer Störung innerhalb eines Bereiches der Funktionsfähigkeit zu bleiben}*
dynamics Dynamik *f*

E

E function 1. E-Funktion *f*; 2. *s.* energy function
E surface E-Fläche *f* *{dreidimensionale Fläche, die die Energiezustände in einem neuralen Netz darstellt}*
economical macromodel ökonomisches Makromodell *n*
economical micromodel ökonomisches Mikromodell *n*
economical model ökonomisches Modell *n*
editor Editor *m* *{Software-Werkzeug zur Änderung von anderer Software}*
educational software Lernprogramm *n*
effect/to bewirken
effect Wirkung *f*, Effekt *m*
effectiveness Effektivität *f*
effector Effektor *m*, Stellelement *n*, Stellorgan *n*
efferent neuron Steuerneuron *n* *{sendet Signale vom Gehirn zu anderen Teilen des Körpers}*
efficiency Effizienz *f*, Wirkungsgrad *m*; *{s.a.* yield*}*
efficiency of an innovation process Effektivität *f* des Innovationsprozesses
element 1. Element *n*, Glied *n* *{eines Systems}*; 2. Übertragungsglied *n*
elementary elementar
elementary alternative Elementaralternative *f*, Vollalternative *f*, Maxterm *m*, Volldisjunktion *f*
elementary assertion elementare Beziehung *f*
elementary automaton Elementarautomat *m*, elementarer Auto-mat *m*
elementary chain elementare Kette *f* *{Graph}*
elementary character Elementarzeichen *n*
elementary conjunction Elementarkonjunktion *f*, Vollkonjunktion *f*, Minterm *m*
elementary cycle elementarer Zyklus *m*, Elementarzyklus *m* *{Graphentheorie}*
elementary decision Elementarentscheidung *f*
elementary event Elementarereignis *n*
elementary fact elementare Beziehung *f*
elementary language elementare Sprache *f*
elementary logical connection elementare logische Verknüpfung *f*, logische Elementarverknüpfung *f*
elementary optimization Elementaroptimierung *f*, Optimierung *f* des Elementarsystems
elementary relationship elementare Beziehung *f*
elementary system elementares System *n*, Elementarsystem *n*
elementary term Elementarterm *m*
eliminability Eliminierbarkeit *f*
eliminable eliminierbar
eliminate/to eliminieren
eliminating induction ausscheidende Induktion *f*
embed/to einbetten *{z.B. eine Computersprache in eine andere}*
embedded expert system eingebettetes Expertensystem *{empfängt Input von Sensoren und gibt Output an Effektoren ab}*

embedded Markov chain eingebettete Markow-Kette f *{Bedienungstheorie}*
empiric[al] formula empirische Formel f
empiric[al] model empirisches Modell n
empiric[al] solution empirische Lösung f
empiric[al] value Erfahrungswert m
empty leer
empty graph leerer Graph m
empty list leere Liste f *{Liste aus null Elementen}*
empty set leere Menge f
empty shell leeres Shell n, leere Schale f *{Expertensystem}*
encryption Codierung f *{zur Geheimhaltung}*, Geheimverschlüsselung f
end edge Endkante f *{Baum}*
end effector Effektor m, Betätigungsorgan n, Handhabungsorgan n *{z.B. Greifer}*
end point 1. Endpunkt m *{eines Vektors}*; 2. Endstellung f; Endwert m; 3. Umschlagpunkt m *{Titration}*
energy function Energiefunktion f *{definiert Konnektion-Matrix und Anfangsinput und beschreibt den Energiezustand des Netzwerkes}*
energy minimum Energieminimum n *{stabiler Zustand der Energiefläche}*
engineering cybernetics technische Kybernetik f
English logic $s.$ non-inverting logic
enqueue/to in die Warteschlange einreihen
entity [begriffliche] Einheit f;

Objekt n *{als Teil einer Klasse von Objekten mit gleichen Eigenschaften, z.B. ein Mensch}*
environment Umgebung f *{z.B. Programmierumgebung}*
environment tool Umgebungsinstrument n
environmental model Umweltmodell n, Modell n der Umgebung (Außenwelt)
equality Gleichheit f
equate/to gleichsetzen
equation Gleichung f
equation solver Gleichungslöser m
equipartition Gleichverteilung f
equipartition property Gleichverteilungseigenschaft f
equipotent gleichmächtig *{Mengenlehre}*
equipotent set gleichmächtige Menge f
equiprobable gleichwahrscheinlich, äquiprobabel
equivalence Äquivalenz f, logische Äquivalenz f *{logische Operation}*
equivalence element Äquivalenzelement n, Äquivalenzglied n, Äquivalenzgatter n, Äquivalenzschaltung f
equivalence operation Äquivalenzoperation f, WENN-UND-NUR-WENN-Operation f
equivalence principle Äquivalenzprinzip n
equivalence relation Äquivalenzrelation f, Äquivalenzbeziehung f
equivalent gleichwertig, äquivalent; von gleichem Wahrheitswert
equivocation Mehrdeutigkeit f
ergodic process ergodischer Prozeß m

ergodic state ergodischer Zustand *m*
ergodic system ergodisches System *n*
ergodic theory Ergodentheorie *f*
error 1. Fehler *m*; Irrtum *m*; 2. Abweichung *f*; 3. Unsicherheit *f*
error backpropagation Fehlerzurückverfolgung *f*
error correctability Fehlerkorrigierbarkeit *f*
error debugging Fehlerbeseitigung *f*, Beseitigung *f* von Fehlern
error diagnostics Fehlerdiagnose *f*
error of the decision element Fehler *m* des Entscheidungselementes
evader Verfolgter *m* *{Verfolgungsspiel}*
evaluate/to auswerten, bewerten, errechnen; abschätzen
evaluate an expression/to den Wert eines [logischen] Ausdruckes berechnen
evaluation Auswertung *f*
evaluation function Bewertungsfunktion *f*, Auswertungsfunktion *f* *{meist heuristische Funktion zur Bewertung der verschiedenen von einem Knoten eines Suchbaumes abgehenden Pfade}*
evaluation net E-Netz *n* *{Petri-Netz}*
evaluation system Auswertungssystem *n*, Bewertungssystem *n*
evaluator Bewerter *m*
evasion condition Verfolgungsbedingung *f*
evasion game Verfolgungsspiel *n*
evasion strategy Verfolgungsstrategie *f*

event Ereignis *n*, Vorgang *m*
event algebra Ereignisalgebra *f*
event-by-event simulation zeittreue Simulation *f*
event compatibility Ereigniskompatibilität *f*
event-driven ereignisgesteuert *{vorwärtsverkettende Problemlösung auf der Basis des jeweils aktuellen Problemstatus}*
event-driven control structure ereignisgesteuerte (datengesteuerte) Steuerstruktur *f* *{Struktur eines Weges zur Problemlösung mit Vorwärtsargumentation aus Anfangsbedingungen}*
event-driven logic ereignisgesteuerte Logik *f*
event field Ereignisfeld *n*
event indicator Indikator *m* eines Ereignisses *{diskrete Verteilung}*
event mark Ereignismarke *f*
event-oriented network ereignisorientierter Netzplan *m*, Vorgangspfeilnetz *n*
event sequence Ereignisfolge *f*
event space Ereignisraum *m*
everyday situation Alltagssituation *f*
evoke/to hervorrufen *{Verhalten}*
example Beispiel *n*
example-driven system beispielgesteuertes System *n*, Induktionssystem *n*
exception Ausnahme *f*
exchangeability Austauschbarkeit *f*, Auswechselbarkeit *f*
excitatory erregend, Erregung übertragend
excitatory synapse erregende Synapse *f* *{bewirkt eine Erhöhung des*

Aktivierungsniveaus des empfangenden Neurons}
excitatory tendency Feuerneigung *f*, Neigung *f* zum Feuern *{des empfangenden Neurons}*
excluded contradiction ausgeschlossener Widerspruch *m*
excluded-third rule Satz *m* vom ausgeschlossenen Dritten, tertium non datur
exclusion Inhibition *f*, Ausschliessung *f*, Exklusion *f*; *{s.a.* NAND, NOT AND, Sheffer function*}*
exclusive exklusiv, ausschließlich
exclusive OR ENTWEDER-ODER *n*, exklusives (ausschließendes) ODER *n*, Antivalenz *f*, Kontravalenz *f*, Disjunktion *f*
exclusive OR circuit (element, gate, network) ausschließende ODER-Schaltung *f*, ausschließendes ODER-Gatter (ODER-Tor) *n*, Antivalenzschaltung *f*, ENTWEDER-ODER-Schaltung *f*, ENTWEDER-ODER-Gatter *n*, ENTWEDER-ODER-Tor *n*
exclusive OR operation ausschließende ODER-Operation *f*, Antivalenz *f*
exhaustive search erschöpfende Suche *f {Suchverfahren, bei dem jeder mögliche Weg eines Baumes oder Netzes durchsucht wird}*
existence-of-infinite-set principle Unendlichkeitsprinzip *n {Mengenlehre}*
existence sentence Existenzaussage *f*
existence theorem Existenzsatz *m*, Existenztheorem *n*
existential proposition Existentialaussage *f {Logik}*
existential quantification Partikularisierung *f*
existential quantifier partikulärer (existentieller) Quantifikator *m*, Existenzquan[tifika]tor *m*, Existenzoperator *m*, Existenzfunktor *m*, Seinszeichen *n*, Einsquantor *m* *{Logik}*
EXOR *s.* exclusive OR
expandable ausdehnbar, erweiterungsfähig; aufrüstbar
expandable structure erweiterbare Struktur *f*
expanded model erweitertes Modell *n*
expectancy Erwartung *f*
expectation Erwartung *f*
expectation-driven erwartungsgesteuert
expectation-driven reasoning erwartungsgesteuerte Schlußfolgerung *f {Problemlösungsmethode, die Hypothesen bezüglich zu erwartender Ereignisse aufstellt}*
expectation-guided solution approach erwartungsgeführter Lösungsweg *m*
experience Erfahrung *f*
experience store Erfahrungsspeicher *m*
experiential knowledge Erfahrungswissen *n*, Oberflächenwissen *n*, heuristisches Wissen *n*
experiment designing (planning) Versuchsplanung *f*
experimental game Planspiel *n*
experimental plan Versuchsplan *m*
experimental result Versuchsergebnis *n*
expert Experte *m*; Expertensystem *n*

expert estimation Expertenschätzung *f*
expert shell Expertensystemshell *n*, Expertensystemschale *f*
expert system Expertensystem *n*
expert tool Expertenwerkzeug *n*
expertise Erfahrung *f*; Wissen *n*, Expertenwissen *n*, Fachwissen *n*
explanation Erklärung *f*, Begründung *f*, Rechtfertigung *f* *{für Handlungsweise}*, Erläuterung *f*
explanation-based learning Lernen *n* durch Erklärung
expression [logischer] Ausdruck *m* *{s.a.* propositional form*}*
extended simplex method erweiterte Simplexmethode *f*
extension 1. Extension *f*, Umfang *m* *{Menge}*; 2. Verlängerung *f*; Dehnung *f*, Ausdehnung *f*
extension field Erweiterungskörper *m*
extensional logic extensionale Logik *f*
extensionality principle Extensionalitätsprinzip *n* *{z.B.* Mengenlehre*}*
external entity externes Objekt *n* *{Quelle oder Senke}*
external input äußerer Input *m* *{Systemtheorie}*
external logic äußere Logik *f*
external model äußeres (externes) Modell *n*
extremal Extremale *f* *{Kurve}*
extremal problem Extremalproblem *n*
extremal system Extremalsystem *n*
extremal value Extremum *n*, Extrem[al]wert *m*
extremize/to extremieren *{minimieren oder maximieren}*

extremizing criterion Extremierungskriterium *n*
extremum Extremum *n*, Extrem[al]wert *m*
extremum-searching adaptive system Extremalsystem *n*
extremum value Extremum *n*, Extrem[al]wert *m*

F

facet Facette *f* *{Feld für Einträge in einem Frame}*
fact Tatsache *f*, Fakt *m* *{Feststellung, deren Gültigkeit allgemein akzeptiert ist, bei Expertensystemen eine bewertete Eigenschaft}*; Fakt *m* *{Paar zusammengehöriger Eingabe- und gewünschte Ausgabemuster, die zum Trainieren eines neuronalen Netzwerkes verwendet werden}*
fair game faires Spiel *m*
fallacy Trugschluß *m* *{Logik}*
false falsch, unwahr *{Logik}*
false conclusion falsche Folgerung (Konklusion) *f*
false decision Fehlentscheidung *f*
false [declarative] sentence falsche Aussage *f* *{Logik}*
fault diagnosis Fehlerdiagnose *f*
fault tolerance Fehlertoleranz *f* *{Fähigkeit eines neuronalen Netzes, die Verarbeitung, eventuell mit verminderter Genauigkeit, fortzuführen}*
fault tree Fehlerbaum *m*
fault tree analysis Fehlerbaumanalyse *f*, deduktive Fehleranalyse *f*
feasibility Zulässigkeit *f*, Durchführbarkeit *f*
feasible brauchbar; zulässig

feasible direction brauchbare (zulässige) Richtung f *{Gradientenmethode}*
feasible-directions method Methode f der zulässigen Richtungen *{nichtlineare Optimierung}*
feasible region zulässiger Bereich m *{z.B. einer Funktion}*
feasible strategy zulässige Strategie f
feature Eigenschaft f, Merkmal n; Umgebungseigenschaft f *{z.B. Farbe oder Tonhöhe}*
feature analysis Merkmalsanalyse f
feature detector Eigenschaftsdetektor m *{Neuronengruppe zur Erkennung einer speziellen Eigenschaft}*
feature filter Merkmalsfilter n
feature map Abbildungsbereich m *{Gehirnbereich, der direkt mit Sensorbereichen des Körpers verbunden ist}*
feature selection Merkmalsselektion f
feature vector Merkmalsvektor m
feedback Rückkopplung f, Rückführung f *{die Ausgangssignale eines Neurons werden an die Eingänge anderer Neuronen derselben oder einer vorangehenden Schicht zurückgeführt; manchmal auch als "Erwartung" bezeichnet}*
feedforward Vorwärtskopplung f *{Neuronen erhalten ihre Inputs nur von der vorangehenden Schicht und geben ihre Outputs nur an die nachfolgende Schicht ab}*
Fibonacci number Fibonacci-Zahl f
Fibonacci search Fibonacci-Suche f
Fibonacci search process Fibonacci-Suchprozeß m, Suchprozeß m mit durch Fibonacci-Zahlen bestimmter Anzahl Tests
fictitious activity Scheinaktivität f, fiktive Aktivität f, fiktiver Vorgang m *{Netzplan}*
fictitious game method Methode f des Scheinspiels *{lineare Optimierung}*
fictive game fiktives Spiel n
fictive player fiktiver Spieler m
field 1. Feld n; 2. Feldgröße f; 3. Körper m *{Mathematik}*; 4. Feld n, Datenfeld n, Eintrag m
field axiom Körperaxiom n
field expression Feldausdruck m *{Arithmetik}*
field extension Körpererweiterung f
field of events Ereignisfeld n
field of lines of behaviour Feld n der Verhaltenslinien
field-programmable logic-sequence feldprogrammierbare logische Folgesteuereinheit f
fifth-generation computer Computer m der fünften Generation *{intelligenter, parallelarbeitender Nicht-von-Neumann-Computer}*
figure 1. Ziffer f; Zahl f; Zahlenwert m; *{s.a. value}*; 2. Abbildung f, Bild n; Zeichen n
figure extraction Merkmalsgewinnung f bei Bildern
final event Endereignis n *{Netzplantechnik}*
final state Endzustand m
fine grain parallelism Neokonnektionismus m
finger Finger m *{Greiferhand am Roboter}*
finite endlich, finit

finite automata theory Theorie *f* der endlichen Automaten
finite automaton endlicher Automat *m*, sequentielle Maschine *f*
finite condition Endzustand *m*
finite-dimensional endlichdimensional
finite-element model Finite-Elemente-Modell *n*
finite game endliches Spiel *n*
finite graph finiter (endlicher) Graph *m*
finite group endliche Gruppe *f*
finite-memory automaton Automat *m* mit endlichem Speicher
finite-memory filter Filter *n* mit begrenztem Gedächtnis
finite-state acceptor endlicher Akzeptor *m*
finite-state machine endlicher Automat *m*, sequentielle Maschine *f* *{s.a.* finite automaton*}*
finite-tree game Spiel *n* mit endlichem Baum
fire/to 1. zünden *{z.B. Thyristor}*; auslösen *{z.B. Horizontalablenkung beim Oszilloskop}*; 2. Maßnahmen durchführen *{die von der rechten Seite einer Regel beschrieben sind}*, [eine Regel] feuern
fire Feuer *n* *{Befehl zum Ausführen einer Regel}*, Zünd-
firing frequency Feuerhäufigkeit *f* *{Maß für die Aktivität eines Neurons, ausgedrückt durch Anzahl Impulse pro Sekunde}*
first-level set Menge *f* erster Ordnung (Stufe)
first-order approximation Näherung (Approximation) *f* erster Ordnung
first-order predicate logic Prädikatenlogik *f* erster Ordnung
first-order set *s.* first-level set
first-order supercharacter Superzeichen *n* erster Ordnung
five-colour theorem Fünffarbensatz *m* *{Graphentheorie}*
fixed-initial state automaton initialer Automat *m*
fixed-level set Menge *f* erster Ordnung (Stufe)
fixed reference model festeingestelltes Bezugsmodell *n*
fixed-sequence robot festprogrammierter Roboter *m*
floating state gleitender Zustand *m*
flow 1. Strömung *f*, Fluß *m*, Durchfluß *m*; 2. Ablauf *m*; 3. Strom *m* *{Graph}*
flow graph Flußgraph *m* *{s.a.* flow diagram*}*
flow line processing Fließbandverarbeitung *f* *{s.a.* pipeline system*}*
follow-the-leader-feedback network FLF-Schaltung *f* *{mehrfach verkoppelte Kaskadenschaltung, bei der alle Rückführungen auf den Eingang zurückgehen}*
follow-the-leader structure Folgestruktur *f*, Follow-the-leader-Struktur *f* *{eine Regelkreisstruktur mit mehrschleifiger Rückführung}*
follow-up node Nachfolgeknoten *m*, nachfolgender Knoten *m*
follow-up state Folgezustand *m*
follower relation Nachfolgerrelation *f*
follower state Nachfolgerzustand *m*
Food's algorithm Foodscher Algorithmus *m* *{Netzplantechnik}*; Foodscher Funktor *m*, Junktor *m* *{Logik}*
forbidden combination verbotene (unzulässige) Kombination *f*

forbidden state verbotener Zustand *m*
forecast[ing] Vorhersage *f*
forest Wald *m* {*Graph*}
forgery Fälschung *f*
form 1. Formblatt *n*, Formular *n*; 2. Ausdruck *m* {*Logik*}
formal formal
formal language formale Sprache *f*
formal logic formale Logik *f*
formal operation formale Operation *f*
formal parameter formaler Parameter *m*
formal statement formale Aussage *f*
formalism Formalismus *m*
formalize/to formalisieren
formalized language formalisierte Sprache *f*
forming rule Bildungsregel *f* {*logische Syntax*}
formula Formel *f*
formula for the predicate calculus prädikatenlogische Formel *f*
formula translator Formelübersetzer *m*
forward vorwärts; Vorwärts-
forward branch Vorwärtszweig *m*
forward chaining Vorwärtsverkettung *f* {*ereignis- oder datengesteuertes Schlußfolgern, eine Strategie*}
four-colour problem Vierfarbenproblem *n* {*Graphentheorie*}
four-person game Vierpersonenspiel *n*
four-valued logic vierwertige Logik *f*
fractal Fraktal *n*

fractel Fraktal *n*
frame Rahmen *m* {*eine Datenstruktur*}, Eigenschaftsliste *f*, Objekt *n*
frame element Rahmenelement *n*
framework Gerüst *n* {*Graph*}
free frei
free automaton freier Automat *m*
free interpretation freie Interpretation *f*, Herbrand-Interpretation *f*
free text retrieval Freitextretrieval *n*
free variable freie Variable *f*
front end Benutzeroberfläche *f*
full voll
full acquisition Volltexterfassung *f*
function 1. Funktion *f*, Aufgabe *f*, Arbeitsweise *f*, Betriebsweise *f*, Funktionsweise *f*; 2. Funktion *f* {*Mathematik*}; 3. Rückkopplungsfunktion *f*
function simulator Funktionssimulator *m*
functional application funktionelle Anwendung *f* {*Problemlösung*}
functor Funktor *m*, Junktor *m*{*Logik*}
fundamental Grundwelle *f*, Grundfrequenz *f*; Fundamental-
fundamental alternative Fundamentalalternative *f*
fundamental approach grundsätzlicher Lösungsweg *m*
fundamental conjunction Fundamentalkonjunktion *f*
fundamental rule Grundgesetz *n*
fundamental symbol Grundsymbol *n* {*z.B. in der Aussagenlogik*}
fundamental term Fundamentalterm *m*
fuzziness Unschärfe *f*

fuzzy unscharf, unexakt, vage; unvollständig; uneindeutig
fuzzy algorithm unscharfer Algorithmus *m*
fuzzy automaton unscharfer Automat *m*
fuzzy distribution unscharfe Verteilung *f*
fuzzy information unscharfe (unexakte) Information *f*
fuzzy logic unscharfe Logik *f*
fuzzy quantor unscharfer Quantor *m*
fuzzy set unscharfe Menge *f* *{Zugehörigkeit der Elemente kann graduell abgestuft sein}*
fuzzy system unscharfes (unexaktes, vages) System *n* *{ein stochastisches System}*

G

gain 1. Verstärkung *f*; Gewinn *m*; 2. Skalierungsfaktor *m* *{Ausgangsgrößenänderung pro Dekade Eingangsgrößenänderung beim logarithmischen Verstärker}*; 3. Übertragungsfaktor *m*
gain function Gewinnfunktion *f*, Auszahlungsfunktion *f* *{Spieltheorie}*
gain matrix Gewinnmatrix *f*, Auszahlungsmatrix *f* *{Spieltheorie}*
gain maximization Gewinnmaximierung *f*, Maximierung *f* des Gewinns
gain optimization Gewinnoptimierung *f*, Optimierung *f* des Gewinns
Galton board Galton-Brett *n*
gamble Glücksspiel *n*

game Spiel *n*, Partie *f* *{z.B. Schach}*
game against nature Spiel *n* gegen die Natur, statistisches Spiel *n*
game behaviour Spielverhalten *n*
game matrix Spielmatrix *f*
game model Spielmodell *n*
game of hazard Hasardspiel *n*, Glücksspiel *n*
game of survival Spiel *n* ums Überleben
game of timing Spiel *n* zur Wahl eines Zeitpunktes
game on a graph Spiel *n* auf einem Graphen
game on the unit square Spiel *n* auf dem Einheitsquadrat
game situation Spielsituation *f*, Partiesituation *f*
game-theoretic spieltheoretisch
game-theoretic model spieltheoretisches Modell *n*
game theory Spieltheorie *f*
game value Spielwert *m*, Wert *m* des Spiels
game without side payment Spiel *n* ohne Nebenauszahlungen
gaming Spielsimulierung *f*
ganglion Ganglion *n* *{Nervenzellen außerhalb des autonomen Nervensystems}*
gate 1. Schaltglied *n*, [logisches] Verknüpfungsglied *n*, Gatter *n*, Tor *n* *{logische Verknüpfungsschaltung}*
gate circuit Torschaltung *f*, Gatterschaltung *f*, logische Verknüpfungsschaltung *f*
general allgemein; verallgemeinert
general algorithm verallgemeinerter Algorithmus *m*
general Erlang distribution allgemeine Erlang-Verteilung *f*

general problem solver allgemeiner Problemlöser *m* {künstliche Intelligenz; Vorgänge der Expertensysteme}
general-purpose robot Universalroboter *m*, Allzweckroboter *m*
general solver *s.* general problem solver
general systems theory allgemeine Systemtheorie *f*
general topology allgemeine [mengentheoretische] Topologie *f*
generality Allgemeingültigkeit *f*, Verallgemeinerung *f*
generalization [rule] Verallgemeinerung *f*, Generalisierung *f* {Prädikatenlogik}
generalize/to generalisieren, verallgemeinern
generalized coordinates verallgemeinerte Koordinaten *fpl*, generalisierte Koordinaten *fpl*
generalized graph verallgemeinerter Graph *m*
generate/to generieren {z.B. Computergraphik}; [Lösung] erstellen; erzeugen {z.B. Kurve}
generating grammar generative Grammatik *f*
generation 1. Generation *f* {z.B. in der technischen Entwicklung}; 2. Erzeugung *f* {z.B. von Programmen}; Synthese *f*
generation-one-point-five robot Roboter *m* der 1,5ten Generation {sensorgesteuert, kann eigenes Arbeitsergebnis bedingt prüfen, Werkzeuge, Werkstücke und Teile erkennen, die vom Sensor erfaßten Daten haben Priorität vor den programmgespeicherten}
generation-three robot Roboter *m* der dritten Generation {intelligent, unterstützt Lösung von firmenspezifischen Problemen}
generation-two robot Roboter *m* der zweiten Generation {erkennt Objekte und verrichtet Handhabungen nach Wechselwirkung zwischen Erkennung und Greiferhand}
generative grammar generische Grammatik *f* {eine formale Grammatik, als eine Menge logischer Schlüsse dargestellt}
generator Generator *m*
generic identifier mnemonischer Kennzeichner *m*, mnemonische Kennung *f*
generic property generische Eigenschaft *f*
generic routine generische Routine *f*, generisches Programm *n*
genetic algorithm genetischer Algorithmus *m*
geometrical information geometrische Information *f*, Weginformation *f* {numerisch gesteuerte Maschine}
gestalt Gestalt *f*, Form *f*, Muster *n* {als Ganzes erkannt}
global umfassend, global
global data base globale Datenbasis *f* {Dateiensammlung zur Beschreibung eines Problems, des Problemstatus und des Lösungsprozesses}
global extreme value globales Extremum *n*
global maximum globales (absolutes) Maximum *n*
global minimum globales (absolutes) Minimum *n*
global optimum globales (absolutes) Optimum *n*

global search globale Suche *f*
global solution globale Lösung *f*
global stability globale Stabilität *f*
global variable globale Variable *f*
glossary Konstantenspeicher *m*, Wörterbuch *n*
gn function Signumfunktion *f*
goal Ziel *n* *{z.B. bei der Problemlösung}*
goal-directed inference zielgesteuerte Inferenz *f*, zielgerichtetes Schließen *n*
goal driving Zielsteuerung *f* *{Problemlösungsmethode, bei der vom Ziel ausgehend rückwärts gegangen wird}*
goal regression Zielregression *f* *{Verwendung von Teilzielen bei der Problemlösung}*
goal-searching system zielsuchendes System *n*, selbsteinstellendes System *n*
goal state Zielzustand *m*
Goedel's theorem of incompleteness Gödelsches Unvollständigkeitstheorem *n*
golden section search procedure Suchverfahren *n* nach dem Goldenen Schnitt
goodness criterion Gütekriterium *n* *{Suchverfahren}*
goodness of fit Güte *f* der Anpassung
gorge search technique Schluchtensuchverfahren *n* *{Optimierung}*
grad grad *{Gradient}*
graded hierarchisch geordnet
gradient Gradient *m* *{maximale Änderungsrate einer Funktion}*
gradient method Gradientenverfahren *n*

gradient projection method Gradientenprojektionsmethode *f*
gradient scheme Steigungsschema *n* *{Methode der dividierten Differenzen}*
gradient search Gradientensuche *f*, Suche *f* nach dem Gradientenverfahren
gradient vector Gradientenvektor *m*
grammar Grammatik *f*
graph Graph *m*, Netzplan *m*; Diagramm *n*, Schaubild *n*
graph structure Graphenstruktur *f*
graph structure representation Graphenstrukturrepräsentation *f*
graph structure transformation Graphenstrukturtransformation *f*
graph technique Netzplantechnik *f*
graph theory Graphentheorie *f*
graphic evaluation and review technique GERT-Methode *f* *{Netzplantechnik}*
graphic language Graphiksprache *f*
graphic representation graphische Darstellung *f*
graphics Graphik *f*, graphische Anzeige *f*; *{s.a. graphic}*
graphics language Graphiksprache *f*
graphics manipulating language graphische Manipulationssprache *f*
group theory Gruppentheorie *f*
grow without bound/to über alle Grenzen wachsen
growing automaton wachsender Automat *m*
growth Wachstum *n*
growth curve Wachstumskurve *f*
growth model Wachstumsmodell *n*

H

habituation Gewöhnung *f* *{abnehmende Empfindlichkeit für einen Reiz, wenn dieser bereits erwartet wird}*
half-group theory Halbgruppentheorie *f*
half-ordered set halbgeordnete Menge *f*
Hamilton line Hamilton-Linie *f* *{Graphentheorie}*
Hamming distance Hamming-Abstand *m* *{Anzahl der Neuronen mit unterschiedlichen Zuständen in zwei Mustern}*
handling machine (robot) Industrieroboter *m*, Handhabungsroboter *m*, Handhabungsautomat *m*
handling system Handhabungssystem *n* *{Roboter}*
handshaking Quittieren, Bestätigen
happen/to sich ereignen, eintreten *{Ereignis}*
hard hart, fest
　hard-and-fast rule Faustregel *f*, Daumenregel *f*
　hard limiter Begrenzer *m*; Sprungfunktion *f*
　hard-wired logic festverdrahtete Logik *f*
hardware 1. Hardware *f*, Geräte *npl*, Bauelemente *npl*, Einrichtungen *npl*; 2. Kleinteile *npl*; Befestigungsmittel *npl*
　hardware level Hardwareebene *f*
hazard Hasard *m*, Wettlauf *m* *{Erscheinung bei logischen Verknüpfungsschaltungen}*
　hazard detection Hasarderkennung *f*
　hazard game Glücksspiel *n*
　hazard method Zufallsmethode *f*
　hazard phenomenon Hasarderscheinung *f*
　hazard problem Hasardproblem *n*
head-of-the-line priority relative Priorität *f* *{Bedienungstheorie}*
heal/to heilen *{z.B. Bauelemente}*
heap Heap *m* *{vollständiger binärer Baum mit hierarchisch abnehmenden Knotenwerten}*
Hebb's rule Hebb-Regel *f* *{eine Lernregel}*
hereditary system Erbsystem *n*, Gedächtnissystem *n*, System *n* mit verzweigten Parametern
hereditary under conclusion folgerungserblich
heuristic heuristisch
heuristic knowledge Oberflächenwissen *n*, Erfahrungswissen *n*, heuristisches Wissen *n*
heuristic method heuristische Methode *f*
heuristic model heuristisches Modell *n*
heuristic programming heuristische Programmierung *f* *{s.a. artificial intelligence}*
heuristic recognition method heuristische Erkennungsmethode *f*
heuristic routine heuristisches Programm *n*
heuristic rule heuristische Regel *f*, Faustregel *f*
heuristic search heuristische Suche *f*, Suche *f* nach Faustregeln *{Suche mittels Testens von Zwischenzuständen entlang möglicher Lösungspfade auf der Basis heuristischen Wissens}*

heuristics Heuristik *f {Lehre von den Problemlösungen nach empirischem Wissen und Erfahrungsregeln}*
hidden element verstecktes (verborgenes) Element *n*
hidden layer verborgene Schicht *f*
hidden line verdeckte Linie *f*
hidden neuron verdecktes (verborgenes) Neuron *n {zwischen Input- und Outputneuronen angeordnet und nur mit diesen verbunden}*
hide/to verbergen *{z.B. inneren Aufbau von Informationsstrukturen}*; unsichtbar machen *{z.B. vorgegebene Daten auf dem Bildschirm}*
hiding Geheimhaltung *f*; verborgene Aufbewahrung *f*
hiding algorithm Algorithmus *m* für verborgene Oberflächen *{Computergraphik}*
hierarchical hierarchisch
hierarchical classification hierarchische Klassifikation *f*
hierarchical computer network hierarchisches Computernetz *n*
hierarchical computer system Computerhierarchiesystem *n*, hierarchisches Computersystem *n*
hierarchical-distributed computer network Computernetzwerk *n* mit hierarchisch verteilter Intelligenz
hierarchical level Hierarchieebene *f*
hierarchical network hierarchisches Netz *n*
hierarchical order hierarchische Ordnung *f*
hierarchical planning hierarchisches Planen *n {Planen zunächst auf höherer Ebene nach den wichtigsten Aspekten, danach zunehmend detailliert auf untergeordneten Ebenen}*
hierarchical structure hierarchische Struktur *f*, Hierarchiestruktur *f*
hierarchical system Hierarchiesystem *n*, hierarchisches System *n*
hierarchy Hierarchie *f {Ordnung in mehreren einander untergeordneten Ebenen}*
hierarchy level Hierarchieebene *f*
hieroglyph Hieroglyphe *f*
high hoch
high-level system hochorganisiertes System *n*
high-order language höhere Computersprache *f*, Sprache *f* höherer Ordnung
high-speed computer schneller Computer *m*
higher-level computer übergeordneter Computer *m*
higher-level estimator Zustandsschätzer *m* in der höheren Ebene *f {hierarchisches System zur Zustandsschätzung}*
higher-level optimization Optimierung *f* auf oberer Ebene *{Mehrebenenoptimierung}*
higher-order approximation Näherung (Approximation) *f* höherer Ordnung
higher-order correction Korrektur *f* höherer Ordnung
higher-order logic, higher predicate calculus Prädikatenlogik *f* höherer Stufen
Hilbert space Hilbert-Raum *m*
hill-climbing method Schrittoptimierungsmethode *f*, Suchschrittmethode *f*, Gradientenmethode *f*; Methode *f* des steilsten Abstiegs; Methode *f* des steilsten Aufstiegs

hillock Hillock *n {Ursprung des Axons am Soma}*
hippocampus Hippocampus *m {Erweiterung des Cortex}*
history of AI Geschichte *f* der künstlichen Intelligenz
hit Fundstelle *f*, gefundene Stelle *f*, Treffer *m*, Hit *m {nach einer Suche}*
hold/to 1. gelten *{Regel, Gesetz}*; 2. enthalten, gespeichert halten; 3. speichern *{z.B. Byte in einem Register}*
holding process Verweilprozeß *m {Bedienungstheorie}*
hologram Hologramm *n {ein als Speicher verwendbares Interferenzbild*
homeostasis Homöostase *f*
homeostat Homöostat *m {relativ einfaches System mit vier Variablen nach Ashby}*
homeostatic mechanism homöostatischer Mechanismus *m*
homeostatic process homöostatischer Prozeß *m*
homeostatic system homöostatisches System *n*
homologous group homologe Gruppe *f*
homomorphic homomorph
homomorphic image homomorphes Bild *n*
homomorphic mapping homomorphe Abbildung *f*
homomorphism Homomorphismus *m*, Homomorphie *f*, topologische Abbildung *f*
homomorphism theorem Homomorphieprinzip *n*
homotopous group homotope Gruppe *f*

Hopfield associator Hopkins-Assoziator *m {assoziatives Speicher-Netzwerk, bei dem infolge unvollständigen Inputs Wege zu benachbarten Energieminima verfolgt werden}*
Horn clause Horn-Klausel *f {eine Anzahl logisch verknüpfter Anweisungen, die zu genau einer Schlußfolgerung führen}*
Horner's scheme Hornersches Schema *n*
household robot Haushaltroboter *m*
human-aided machine translation Maschinenübersetzung *f* mit menschlicher Unterstützung
human brain menschliches Gehirn *n*
human engineering Anthropotechnik *f*
human information processing menschliche Informationsverarbeitung *f*
human intelligence menschliche Intelligenz *f*
human interface Schnittstelle *f* zum Menschen
human intervention Eingriff *m* durch den Menschen
human knowledge Wissen *n* des Menschen
human operator Mensch *m* als Bediener (Operateur)
humanlike finger menschenähnlicher Finger *m {am Roboter}*
hyperactive hyperaktiv, überaktiv
hypergraph Hypergraph *m*
hypermatrix Hypermatrix *f*, Übermatrix *f*
hyperplane Hyperebene *f*
hyperspace Hyperraum *m*, Überraum *m*

hypersurface Hyperfläche *f*, Oberfläche *f*
hypoactive hypoaktiv, unteraktiv
hypothesis Hypothese *f*
hypothetic automaton hypothetischer Automat *m*, hypothetische Maschine *f*

I

idea Idee *f*
idea processing Ideenverarbeitung *f*, Verarbeitung *f* von Gedanken
ideal observer idealer Beobachter *m*
idealization Idealisierung *f*
idealize/to idealisieren
idealized model idealisiertes Modell *n*
idealized situation Idealfall *m*
idealized system idealisiertes System *n*
identification problem Identifizierungsproblem *n*
identification process modelling Modellierung *f* des Identifizierungsprozesses
identification system Identifikationssystem *n*, Erkennungssystem *n*
identifier Identifikator *m*, Bezeichner *m*, Kenner *m*, Name *m*, Kennung *f*
identifying algorithm Erkennungsalgorithmus *m*
identifying system Erkennungssystem *n*
identity Identität *f* {*z.B. zwischen zwei logischen Aussagen*}
identity axiom Identitätsaxiom *n*
identity circuit Identitätsschaltung *f*, Identitätsgatter *n*, Identitätselement *n*, Identitätsglied *n*

identity function Identitätsfunktion *f*
identity matrix Identitätsmatrix *f*
identity rule Satz *m* von der Identität, Identitätssatz *m*
IF AND ONLY IF DANN UND NUR DANN, GENAU DANN [WENN] {*Logik*}
IF-THEN WENN DANN {*Wahrheitsfunktion*}
IF-THEN rule Produktionsregel *f* {*künstliche Intelligenz*}
IF-THEN system WENN-DANN-System *n* {*System mit determiniert aufeinanderfolgenden Zuständen*}
illegal unzulässig
illegal character unzulässiges Zeichen *n*
illegal region unerlaubter Bereich *m* {*Suche*}
illegal sequence unzulässiger Ablauf *m*, unzulässige (verbotene) Folge *f* {*z.B. bei sequentiellen Schaltsystemen*}
image 1. Abbildung *f*, Bild *n* {*Mathematik*}; 2. Schirmbild *n*, Bild *n*
image acquisition Bilderfassung *f*
image compression Bildkompression *f*
image degradation Bildverschlechterung *f*
image recognition Bilderkennung *f*
image synthesis Bildsynthese *f*
image understanding Bildverständnis *n* {*Bilderkennung durch Computer einschließlich kognitiver Weiterverarbeitung zur Szeneninterpretation aus Bilddaten*}
imaginary exercise Gedankenversuch *m*

imitate/to imitieren, nachbilden
imitation game Imitationsspiel *n*
impact of reliability in expert systems Einfluß *m* der Zuverlässigkeit bei Expertensystemen
imperative statement unbedingte Anweisung *f*
implementation environment Implementierungsumgebung *f*
implication Implikation *f*, WENN-DANN-Verknüpfung *f*, Produktionsregel *f*
imply/to [dasselbe] bewirken; implizieren, WENN-DANN-verknüpfen
impossibility Unmöglichkeit *f*
impossible event unmögliches Ereignis *n* {Wahrscheinlichkeitstheorie}
incertainty Ungewißheit *f*
incertainty interval Unsicherheitsintervall *n*
incidence function Inzidenzfunktion *f* {Graphentheorie}
inclusion Inklusion *f*, Implikation *f* {Boolesche Verknüpfung}
inclusion relation Inklusionsrelation *f*, Enthaltenseinsrelation *f*
inclusive disjunction *s.* inclusive OR
inclusive OR ODER-AUCH *n*, inklusives (einschließendes) ODER *n*, Disjunktion *f*
inclusive OR element Inklusiv-ODER-Element *n*, inklusives ODER-Element *n*
inclusive OR function einschließende ODER-Funktion *f*, Alternative *f*
incompatible events unvereinbare Ereignisse *npl*
incomplete induction unvollständige Induktion *f*
incomplete information unvollständige Information *f*
incomplete-information game Spiel *n* mit unvollständiger Information
incomplete-information problem Problem *n* mit unvollständiger Information
incomplete statement unvollständige Aussage *f*
incompleteness theorem Unvollständigkeitssatz *m*
inconsistent unvereinbar
inconsistent events unvereinbare (konträre) Ereignisse *npl*
inconsistent system überbestimmtes System *n*
indecidability Unentscheidbarkeit *f*
indecidable unentscheidbar
indecomposable unzerlegbar {logischer Ausdruck}
indecomposable in propositional logic aussagenlogisch unzerlegbar
independent events [voneinander] unabhängige Ereignisse *npl*
indeterminacy Indeterminiertheit *f*, Unbestimmtheit *f*
indeterminacy of strategy Strategieunbestimmtheit *f*
indeterminate unbestimmt
indetermined automaton indeterminierter Automat *m*
indeterministic system indeterministisches System *n*; {s.a. probabilistic} system
indexed indiziert
indexed family of sets durch Index gekennzeichnetes Mengensystem *n*
indifference Indifferenz *f*, Unbestimmtheit *f*
indifferent situation indifferente (unbestimmte) Situation *f*

indistinguishable 54

indifferent stability unbestimmte Stabilität *f*, unbestimmtes Gleichgewicht *n*
indistinguishable nicht unterscheidbar
individual Individuum *n {nicht weiter logisch unterteilbares Element, Atom}*
individual constant Individualkonstante *f {Prädikatenlogik}*
individual variable Individuenvariable *f {Prädikatenlogik}*
induction Induktion *f*, induktiver Schluß *m {Logik}*
induction system Expertensystem *n* mit Beispielen als Wissensbasis *{künstliche Intelligenz}*, Induktionssystem *n*, beispielgesteuertes System *n*
inductive induktiv
inductive conclusion induktive Schlußfolgerung *f*, induktiver Schluß *m*, Induktionsschluß *m*
inductive defined set induktiv definierte Menge *f*
inductive fault analysis induktive Fehleranalyse *f*, Fehleranalyse *f* durch Induktion (induktiven Schluß)
inductive inference induktive Schlußfolgerung *f*
inductive learning induktives Lernen *n {Schließen aus gegebenen Fakten mittels allgemeiner Hypothesen}*
inductive shell induktives (vorwärtsverkettetes) Shell *n {Expertensystem}*
industrial robot Industrieroboter *m*, [industrieller] Roboter *m*, Handhabungsautomat *m*, Manipulator *m*
inexactness Ungenauigkeit *f*

infer/to [logisch] schließen, aus Beweisen folgern; aus Prämissen folgern
inference [logischer] Schluß *m*, Folgerung *f*, Folgerungsoperation *f*
inference chain Schlußkette *f {Logik}*
inference engine Inferenzmaschine *f*, Inferenzmechanismus *m*, Schlußfolgerungssystem *n {Steuerstruktur eines KI-Problemlösers, bei der Steuerung und Wissen voneinander getrennt sind}*
inference machine Folgerungscomputer *m*
inference net Folgerungsnetz *n {Gesamtheit der Schlußketten eines Systems einfacher logischer Verknüpfungen}*
inference procedure Schlußfolgerungsvorgang *m*
inference relation Schlußrelation *f*
inference rule Inferenzregel *f*, Schlußregel *f*
inference strategy Folgerungsstrategie *f*
infimum Infimum *n*
infinite unendlich, unbegrenzt, infinit
infinite automaton unendlicher Automat *m*
infinite-dimensional unendlich-dimensional
infinite game unendliches Spiel *n*
infinite graph unfiniter (unendlicher) Graph *m*
infinite group unendliche Gruppe *f*
infinite machine unendlicher Automat *m*
infinite-tree game Spiel *n* mit unendlichem Baum

infinite-valued logic unendlich wertige Logik *f*
infinity Unendlichkeit *f*
influence Einfluß *m* *{einer Größe auf eine Bezugsgröße}*
information Information *f*; Informationsmenge *f*
information acquisition Informationsaufnahme *f*
information block Informationsblock *m*
information chain Informationskette *f*
information content Informationsgehalt *m*; Informationsinhalt *m*
information content of a feature Informationsgehalt *m* eines Merkmals
information content of a language Informationsmaß *n* einer Sprache
information conversion process Informationsumformungsprozess *m*
information density Informationsdichte *f*
information drain Informationssenke *f*
information flow Informationsfluß *m*, Informationsstrom *m* *{z.B. als Signal im Regelkreis}*
information gathering Informationsgewinnung *f*
information hiding Geheimhaltung *f* von Information
information interchange Informationsaustausch *m*
information lack Informationsmangel *m*, Informationsdefizit *n*
information pick-up Informationsaufnahme *f*
information receiver Informationsempfänger *m*, Empfänger *m*
information reduction Informationsreduktion *f*, Informationsverdichtung *f*
information representation Informationsdarstellung *f*
information retrieval Informationswiedergewinnung *f*, Wiederauffinden *n* von Informationen
information retrieval system Informationsrecherchesystem *n*
information sink Informationssenke *f*
information source Informationsquelle *f*
information space Informationsraum *m*
information structure Informationsstruktur *f*
information supply Informationsvorrat *m*
information system Informationssystem *n*
information theory Informationstheorie *f*
information utilization Informationsnutzung *f*
informational informationell
informational redundancy informationelle Redundanz *f*, Informationsredundanz *f*
inherent inhärent, innewohnend
inherent entropy Eigenentropie *f*
inherent instability natürliche Instabilität *f*
inherent stability Eigenstabilität *f*
inherit/to vererben, erben *{z.B. von Frame zu Frame}*
inheritance Vererbung *f* *{Übertragung der Eigenschaften einer Oberklasse an Unterklassen}*
inheritance hierarchy Vererbungs-

inherited 56

hierarchie *f*, [erbliche] Eigenschaftshierarchie *f*
inherited error mitgeschleppter Fehler *m*
inhibit/to hemmen
inhibitory hemmend, verhindernd
inhibitory gate NAND-Gatter *n*, JEDOCH-NICHT-Gatter *n*, UND-NICHT-Tor *n*
inhibitory synapse entregende Synapse *f* {bewirkt eine Senkung des Aktivitätsniveaus des empfangenden Neurons}
initial anfänglich, initial; Anfangs-, Erst-
initial automaton initialer Automat *m*
initial behaviour Anfangsverhalten *n*
initial concept Anfangskonzept *n*
initial condition Anfangsbedingung *f*, Anfangszustand *m*
initial set Initialmenge *f*
initial-value problem Anfangswertproblem *n*, Cauchysches Problem *n*
injection Injektion *f* {Abbildung in der Topologie}
injective function Injektion *f*, eindeutige Funktion *f* {Logik}
innovation Innovation *f*
innovation representation Innovationsdarstellung *f*
innovation strategies Innovationsstrategien *fpl*
innovative computer architecture innovative Computerarchitektur *f*
input 1. Input *m*, Erregung *f*, Stimulus *m*; 2. Eingangssignal *n*; Eingangsgröße *f*; 3. Eingabewert *m*
input alphabet Eingabealphabet *n*

input logic Eingangslogik *f*
input neuron Inputneuron *n*, Eingabeneuron *n* {empfängt Daten von der Außenwelt}
input pattern Muster *n* von Eingabedaten
input primitive Eingabeelement *n* {Logik}
input resolution Eingabeauflösung *f* {numerische Steuerung}
input valency Eingangsvalenz *f* {Graph}
inquire/to abfragen, fragen
inquiry Abfrage *f*, Anfrage *f*
inquiry/response Frage/Antwort *f*
instabilitiy Instabilität *f*
instable instabil
instable state instabiler Zustand *m*
instance Beispiel *n*
instantation Fallbetrachtung *f* {Einsetzen eines speziellen Falles für eine oder mehrere Variable}
instantiation Instantiation *f* {Kombination einer Regel und eines Elements einer Faktendatenbank}
instantiation Instanziierung *f*, Einzelfall *m*, Beispiel *n*
instruct/to belehren {Lerntheorie}; lehren {lernender Automat}
instruction Befehl *m*, Maschinenbefehl *m*; Operation *f*
instruction-obeying machine befehlbefolgender Automat *m*
instruction set Instruktionsmenge *f*, Befehlsmenge *f*, Befehlssatz *m*, Befehlsliste *f*, Befehlsvorrat *m*
instruction structure Instruktionsstruktur *f*, Instruktionsaufbau *m*, Befehlsstruktur *f*, Befehlsaufbau *m*

instructional designer Lehrprogrammentwickler *m*
instructional software Ausbildungs-Software *f*, Lern-Software *f*
intelligence Intelligenz *f* *{Fähigkeit eines Individuums oder einer Menge von Individuen, auf Probleme erfolgreich zu reagieren und unter Ausnutzung des eigenen Wissens sowie zusätzlich erhaltener Informationen logisch zu schließen}*
intelligence amplifier Intelligenzverstärker *m*
intelligent intelligent
intelligent assistant program intelligentes Assistenzprogramm *n* *{Expertensystem zur Unterstützung des Menschen bei der Lösung einer Aufgabe}*
intelligent character recognition intelligente Zeichenerkennung *f*
intelligent computer-aided instruction intelligente computerunterstützte Unterweisung *f*
intelligent device intelligentes Gerät *n*
intelligent job aid intelligente Arbeitshilfe *f*
intelligent knowledge-based system intelligentes Wissensbasissystem *n*
intelligent measuring instrument intelligentes Meßgerät (Meßinstrument) *n*
intelligent robot intelligenter Roboter *m* *{frei programmierbar und Anpassung des Bewegungsmusters durch sensorische Wahrnehmungsfähigkeiten}*
intelligent tutoring system intelligentes Lehrsystem *n*

intelligent workstation intelligenter (computergestützter) Arbeitsplatz *m*
intelligibility Erkennbarkeit *f*, Verständlichkeit *f*, Zeichenerkennbarkeit *f*, Signalerkennbarkeit *f*
intensional logic intensionale Logik *f*
inter-arrival time Zwischenankunftszeit *f* *{Bedienungstheorie}*
interaction gegenseitige Beeinflussung *f*, Wechselwirkung *f*, Kopplung *f*
interaction of man and computer Interaktion *f* des Menschen mit dem Computer
interactive interaktiv; Dialog-
interactive application interaktive Anwendung *f*
interactive computing system Computersystem *n* mit direkter Wechselwirkung zwischen Mensch und Maschine, Dialogcomputersystem *n*
interactive environment interaktive Umgebung *f* *{System, bei dem der Benutzer mit dem Computer einen Dialog führt}*
interactive graphics interaktive Graphik *f*
interactive mode Dialogbetrieb *m*
interchange Austausch *m*
interconnected structures vermaschte (vernetzte) Strukturen *fpl*
interconnection Verbindung *f*, Querverbindung *f* *{zwischen Neuronen}*
interdependence wechselseitiger Zusammenhang *m*, gegenseitige Abhängigkeit *f*, Interdependenz *f*
interface Interface *n*, Schnittstelle *f* *{Verbindungsstelle oder Verbin-*

dungsglied zwischen zwei untereinander kommunizierenden [maschinellen oder menschlichen] Komponenten oder Systemen}
interior penalty method Barrieremethode *f {Straffunktionsmethode}*
interior point innerer Punkt *m {Topologie}*
interleave Überlappung *f*, Antwortüberlappung *f*
interleave mode Parallelbetrieb *m*
interlink/to verketten, verbinden *{Systeme}*
interlinking Verkettung *f*, Verbindung *f {von Systemen}*
interlock/to sperren, blockieren, verblocken, verriegeln, abriegeln *{z.B. Signalfluß}*
interlocked recursion verschachtelte Rekursion *f*
intermediate Zwischen-
intermediate field Zwischenkörper *m {Mathematik}*
intermediate language Zwischensprache *f*
intermediate state Zwischenzustand *m*
internal 1. innerlich, intern, im Innern; 2. eingebaut
internal configuration Konfiguration *f* einer programmierbaren Maschine
internal input innere Eingabe *f {Systemtheorie}*
internal model internes (inneres) Modell *n*
internal-model principle Prinzip *n* des inneren Modells
internal-model system System *n* mit internem Modell
internal output innere Ausgabe *f*

{Systemtheorie}
internal penalty function innere Straffunktion *f*
internal procedure interne Prozedur *f*
internal representation interne Darstellung *f*
internal-stable set innerlich stabile Menge *f*
international phonetic alphabet internationales Phonetikalphabet *n*
interpolative network Interpolationsnetzwerk *n {gewinnt aus verrauschten Daten "bessere" Daten}*
interpret/to interpretieren, auswerten, bewerten
interpretation Interpretation *f {z.B. Prädikatenlogik}*
interpretative machine interpretative Maschine *f*
interpretative translation program interpretatives Übersetzungsprogramm *n*
interpreter 1. Zuordner *m {Information}*; 2. Übersetzer *m*; Interpretierer *m*, Interpreter *m*
interrelation Koppelbeziehung *f*
interrogate/to abfragen
interrogation Abfrage *f*
interrogation system Abfragesystem *n*
interruptability Unterbrechbarkeit *f*
interruptable unterbrechbar
interruption Unterbrechung *f*
intersection Durchschnitt *m {Mengenlehre}*
intersection set Durchschnittsmenge *f*
intrinsic instability innere Instabilität *f*, Eigeninstabilität *f*

introspection Selbstbeobachtung *f*
{*Mensch*}
invalid ungültig, unzulässig
invariance Invarianz *f*
invariance theorem Invarianztheorem *n*
inventory model Lagerhaltungsmodell *n*
inverse umgekehrt, entgegengerichtet, invers, reziprok
inverse element inverses Element *n*
inversion Inversion *f*, Umkehrung *f*
{*z.B. in der Logik*}
inversion circuit NICHT-UND-Schaltung *f*
invert/to umkehren, invertieren, negieren
invert gate invertierendes Gatter *n*, NICHT-Gatter *n*, NICHT-Element *n*, NICHT-Glied *n*, Inverter *m*
inverted OR NOR *n* {*logische Funktion*}
inverted-tree dictionary nach dem Prinzip des umgekehrten Baumes organisiertes Adreßbuch *n*
invertibility Invertierbarkeit *f*, Umkehrbarkeit *f*
invertible invertierbar, umkehrbar
invertible function invertierbare (umkehrbare) Funktion *f*, Injektion *f* {*Logik*}
invokation Aufruf *m*
invoke/to aufrufen, in Aktion versetzen
irreflexivity Irreflexivität *f* {*Logik*}
irrelevant information irrelevante Information *f*
isomorphic group isomorphe Gruppe *f*
isomorphism Isomorphismus *m*, bijektiver Homomorphismus *m*
isomorphism of automata Isomorphismus *m* von Automaten
isomorphism of order Ordnungsisomorphie *f*
isomorphism problem of dynamic systems Isomorphieproblem *n* dynamischer Systeme
isomorphous isomorph
isoperimetric problem isoperimetrisches Problem *n*
IU *s.* image understanding

J

job Arbeitsaufgabe *f*; Aufgabe *f*; Auftrag *m*; Job *m* {*Computer*}
job aid Arbeitshilfe *f*, Ausführungshilfe *f*
job queue Job-Wartschlange *f*
joining robot Fügeroboter *m*, Roboter *m* zum Fügen von Werkstücken *npl*
joint 1. Verbindung *f*, 2. Verbindungsstück *n*; 3. Gelenk *n*; 4. Fuge *f* {*zwischen benachbarten Werkstücken*}
joint entropy gemeinsame Entropie *f* [an Kanaleingang und Kanalausgang]
joint probability gemeinsame Wahrscheinlichkeit *f*, Verbundwahrscheinlichkeit *f*
joint probability distribution function Verteilungsdichte *f* der Verbundwahrscheinlichkeit {*von zwei stochastischen Variablen*}
jointed-arm robot Gelenkarmroboter *m*, Roboter *m* mit Gelenkarm
judge/to beurteilen
judgeable beurteilbar

K

Karnaugh map Karnaugh-Plan *m*, K-Plan *m*
KBS *s.* knowledge-based system
key phrase Schlüsseltext *m*, Keytext *m* *{Textbaustein bei der Textverarbeitung}*
keyboard/to eintasten, eintippen; erfassen *{Text}*
keytext Schlüsseltext *m*, Keytext *m* *{Textbaustein bei der Textverarbeitung}*
keyword Codewort *n*, Schlüsselwort *n*; Deskriptor *m*, Stichwort *n*
keyword assignment Schlüsselwortzuordnung *f*
keyword-in-context register nach signifikanten Titelworten geordnetes Register *n*
keyword-out-of-context register KWOC-Register *n* *{automatisch aufgestelltes Register, bei dem der Kontext dem Schlüsselwort auf einer nachfolgenden Zeile folgt}*
knapsack problem Knapsackproblem *n*, Rucksackproblem *n*
knowledge Wissen *n*
knowledge acquisition Wissenserfassung *f*, Wissenserwerb *m*, Wissensakquisition *f*
knowledge availability Wissensverfügbarkeit *f*
knowledge base Wissensbasis *f* *{Sammlung von Fakten und Informationen zur Problemlösung}*
knowledge-base chaining Verkettung *f* von Wissensbasen
knowledge base management Verwaltung *f* der Wissensbasis
knowledge-based control Informationsbanksteuerung *f*, wissensbasierte Steuerung *f*
knowledge-based system wissensbasiertes System *n*
knowledge engineering Wissenstechnik *f*, KI-Technik *f* *{Anwendung von Wissen, z.B. in Expertensystemen}*
knowledge information processing system Wissensverarbeitungssystem *n*, Expertensystem *n*
knowledge level Kenntnisstand *m*, Wissensstand *m*
knowledge machine Computer *m* für Expertensysteme
knowledge processing Wissensverarbeitung *f*
knowledge relevance Wissensrelevanz *f*
knowledge representation Wissensdarstellung *f* *{aus der Datenstruktur zur Darstellung des für ein Problem erforderlichen Wissens}*
knowledge sink Wissenssenke *f*
knowledge source Wissensquelle *f* *{Teil des Expertensystems, das das Wissen für eine bestimmte Aktivität oder einen bestimmten Arbeitsbereich liefert}*

L

labyrinth automaton Labyrinthautomat *m* *{Bildauswertung}*
lack of knowledge Mangel *m* an Wissen
language Sprache *f* *{im weitesten Sinn eine Gruppe von Zeichen, Vereinbarungen und Regeln zum Befördern von Nachrichten unter den Aspekten Pragmatik, Semantik und Syntax}*
language architecture Spracharchitektur *f*
language interpreter Sprachinterpreter *m*
language of the predicate calculus prädikatenlogische Sprache *f*
language processing *s.* natural language processing
language-processing computer sprachverarbeitender Computer *m*
language redundancy Sprachredundanz *f*
language translator Sprachübersetzer *m* *{Assembler, Compiler oder anderes Programm zum Übersetzen von Befehlen oder Daten aus einer Sprache in eine andere}*
lateral feedback laterale Rückführung *f* *{Rückführung in einem mehrschichtigen neuronalen Netzwerk in der Form, daß der Input eines Neurons vom Output anderer Neuronen derselben Ebene abhängt}*
law Gesetz *n*
law of chance Zufallsgesetz *n*
layer Schicht *f* *{Gruppe von untereinander verbundenen Neuronen, die an einer bestimmten Funktion beteiligt sind}*

layout optimization Grundrißoptimierung *f*
ld-chart character Beliebigzeichen *n*
leaf Blatt *n* *{Phasenebene}*
leaf node Blattknoten *m*, Terminalknoten *m*, letzter Knoten *m* *{Suchbaum}*
leapfrog structure Bocksprungstruktur *f*, Leap-frog-Struktur *f*, LF-Struktur *f* *{eine Regelkreisstruktur mit mehrschleifiger Rückführung}*
learn pattern Lernmuster *n*
learnability Lernfähigkeit *f*
learning Lernen *n* *{ein neuronales Netzwerk ändert seine Gewichte entsprechend dem Input von außen}*
learning algorithm Lernalgorithmus *m*
learning automaton lernender Automat *m*, lernende Maschine *f*
learning behaviour Lernverhalten *n*
learning by comprehending Lernen *n* durch Erfassen (Begreifen)
learning by conditional allocation Lernen *n* durch bedingte Zuordnung
learning by copying Lernen *n* durch Nachahmung
learning by doing Lernen *n* durch Verrichten
learning by imitating Lernen *n* durch Nachahmung
learning by instruction Lernen *n* durch Belehrung
learning by optimizing Lernen *n* durch Optimieren
learning by storing Lernen *n* durch Speichern
learning by success Lernen *n* durch Erfolg

learning-by-success automaton probierender Automat *m*
learning by understanding Lernen *n* durch Erfassen (Begreifen)
learning classifier Lernklassifikator *m*, lernender Klassifikator *m*
learning computer lernender Computer *m*
learning connection Lernverbindung *f*
learning control lernende Regelung *f*
learning element lernfähiges Element *n*, Lernelement *n*
learning machine lernende Maschine *f*, lernender Automat *m*
learning matrix Lernmatrix *f*
learning mechanism Lernmechanismus *m*
learning model Lernmodell *n*
learning-model technique Lernmodelltechnik *f*
learning phase Lernphase *f*
learning process Lernprozeß *m*
learning program Lernprogramm *n*
learning rate Lernrate *f* {*Faktor zur maßstäblichen Veränderung aller Korrekturen beim Lernprozeß, um die Konvergenzgeschwindigkeit des Netzwerkes zu verbessern*}
learning rule Lernregel *f* {*eine Hebb-Regel, die dem Netzwerk angibt, wie die Werte der Connection-Stärken zu verändern sind*}
learning scheme Lernschema *n*
learning set Lernmenge *f*
learning structure Lernstruktur *f*
learning system lernendes System *n*, Lernsystem *n*
learning-system theory Theorie *f* lernender Systeme
learning theory Lerntheorie *f*
learning with steady-state conditions lineares Lernen *n*, Lernen *n* in stationärer Umgebung
learning without steady-state conditions nichtlineares Lernen *n*, Lernen *n* in nichtstationärer Umgebung, Lernen *n* unter dem Einfluß von Störgrößen
least commitment weitestmögliche Entscheidungsverschiebung *f*, Verzögerung *f* von Entscheidungen zur Problemlösung [bis möglichst viel Informationen vorliegen]
left-hand side Bedingungsteil *m* [einer Regel]
legged locomotion Fortbewegung *f* auf Beinen {*Robotertechnik*}
legible lesbar
legitimate gültig, zulässig
lemma Lemma *n*, Hilfssatz *m*
level 1. Niveau *n*; 2. *s.* hierarchical level; 3. Signalpegel *m*, Pegel *m*
level calculus Stufenkalkül *m* {*Prädikatenlogik*}
level change 1. Sprung *m* {*z.B. Spannungssprung*}; 2. Pegelwechsel *m* {*logische Pegel*}
level of priority Prioritätsebene *f*
lexical analyzer lexikalischer Analysator *m*
lexicographic code lexikographischer Code *m*
likelihood Likelihood *f*
limit of stability Stabilitätsgrenze *f*
limited-sequence robot Roboter *m* mit beschränkter Bewegungsfolge
linear linear; verzerrungsfrei
linear learning lineares Lernen *n*
linear vector space linearer Vektorraum *m*

linearity range Linearitätsbereich *m*
linearized model linearisiertes Modell *n*
link/to verknüpfen *{z.B. Graphen}*; verketten *{z.B. Unterprogramme}*
link 1. Verknüpfung *f*, Bindung *f*; 2. Verbindungselement *n*, Bindeglied *n*; 3. Übertragungsglied *n*; Glied *n* *{in einem semantischen Netz}*; 4. Verbindungsangabe *f*, Verbinder *m* *{z.B. Zeiger}*
link system Zwischenverbindungssystem *n*, Linksystem *n* *{Bedienungstheorie}*
linkage Gelenk *n* *{Robotik}*
linkage operation Verkettungsoperation *f*
linked gekoppelt, verkettet
LIPS *s.* logical inference step per second
list Liste *f* *{die Elemente einer Liste können unteilbar sein oder sind selbst wiederum Listen}*
list processing Listenverarbeitung *f*
literal Literal *n*, Direktoperand *m* *{Atom oder Nichtatom}*
local-coupled filter Filter *n* mit lokaler Kopplung *{zellularer Automat}*
local extremum lokales (örtliches, relatives) Extremum *n*
local minimum lokales (örtliches, relatives) Minimum *n*
local optimum lokales (örtliches, relatives) Optimum *n*
local stability lokale (örtliche) Stabilität *f*
local variable lokale Variable *f*
localization Lokalisation *f* *{tritt auf, wenn eine Anzahl dicht benachbarter Neuronen eine Signalmenge parallel empfängt und als eine Einheit betrachtet}*
logic Logik *f*
logic algebra Algebra *f* der Logik
logic analysis Logikanalyse *f*
logic analyzer Logikanalysator *m*
logic array logisches Feld *n*, Logikanordnung *f*
logic circuit logische Schaltung *f*, Logikschaltung *f*, logische Verknüpfung *f*
logic circuit analyzer Logikanalysator *m*
logic diagram Logikdiagramm *n*, Logikplan *m*, logisches Diagramm *n* *{Schema}*
logic element logisches Verknüpfungselement (Verknüpfungsglied, Element, Glied, Schaltelement) *n*, Logikelement *n*
logical adress logische Adresse *f*
logical AND logische UND-Verknüpfung *f*, Koinzidenz *f*
logical automaton logischer Automat *m*
logical comparison logischer Vergleich *m*
logical connective logischer Operator *m*, logische Konstante *f* *{zur Verknüpfung logischer Aussagen derart, daß der Wahrheitswert der Aussagen den Wahrheitswert der Verknüpfung bestimmt, z.B. UND, ODER}*
logical consequent logischer Schluß *m*
logical contradiction logischer Widerspruch *m*
logical decision logische Entscheidung *f*

logical elementary function logische Elementarfunktion *f*, atomare Funktion *f*
logical equation logische Gleichung *f*
logical equivalence logische Gleichwertigkeit *f*
logical expression logischer Ausdruck *m*
logical inference logische Folgerung *f*, logischer Schluß *m*, maschinelle Deduktion *f*
logical inference step per second logischer Folgerungsschritt *m* pro Sekunde *{Einheit für die Rechenleistung eines Computers bei logischer Programmierung}*
logical instruction logischer Befehl *m*
logical interconnection logische Verknüpfung *f*
logical machine logischer Automat *m*
logical-mathematical calculus logisch-mathematischer Kalkül *m*
logical-mathematical language logisch-mathematische Sprache *f*
logical matrix logische Matrix *f*
logical node logische Verbindungsstelle *f*, logischer Knoten *m*
logical operation logische Operation (Verknüpfung) *f*, Logikoperation *f*
logical representation Wissensdarstellung *f* mittels logischer Verknüpfungen
logical semantics logische Semantik *f*
logical sum logische Summe *f*
logical variable Logikvariable *f*
logician Logiker *m*

logistics Logistik *f*
long-term memory Langzeitgedächtnis *n*
loop/to 1. den Kreis schließen; 2. mit Schlingen versehen *{z.B. Graph}*; 3. Schleife (Programmschleife) fahren
looped graph Graph *m* mit Schleifen
loopless graph Graph *m* ohne Schleifen
loss function Verlustfunktion *f*
loss probability Verlustwahrscheinlichkeit *f*
loss process Verlustprozeß *m* *{Bedienungstheorie}*
loss system Verlustsystem *n*
loss traffic Verlustverkehr *m*, Restverkehr *m*
lower unterer, untere, unteres; Nieder-, Klein-
lower-level estimator Zustandsschätzer *m* in der unteren Ebene *f* *{hierarchisches System zur Zustandsschätzung}*
lower-level optimization Optimierung *f* auf unterer Ebene *{Mehrebenenoptimierung}*
LSH *s.* left-hand side

M

machine Maschine *f*, Automat *m*, Computer *m*; *{s.a.* automaton, computer*}*
machine-aided cognition computergestütztes Erkennen *n*
machine-aided translation of languages maschinengestützte Übersetzung *f*
machine analysis automatische Sprachanalyse *f*

machine dictionary Maschinenwörterbuch *n*, Computerwörter- buch *n*
machine information structure Informationsstruktur *f* einer Computerarchitektur
machine learning Lernen *n* des Computers *{automatische Extraktion von Wissen aus Daten}*
machine logic Computerlogik *f* *{eingebaute Methoden der Problemlösung, Aktivitäten des Computers, erforderliches Datenformat und Art der Entscheidungsfindung}*
machine music *s.* computer-composed music
machine recognition of speech maschinelle Spracherkennung *f*
machine translation automatische (computerisierte) Übersetzung *f*, Maschinenübersetzung *f*
machine translation of languages maschinelle Sprachübersetzung *f*
machine vision maschinelle Bilderkennung *f {z.B. mittels optischer Einrichtung am Industrieroboter}*; Bilderkennung *f* mit Computer *{wissensgestützter und erwartungsgesteuerter Prozeß zur Interpretation von Modellen aus Sensordaten}*
macro evaluation net Makro-E-Netz *n {Petri-Netz}*
macro location Makrostelle *f {Makro-E-Netz}*
magnitude Wert *m {Betrag eines Vektors, gebildet von der Quadratwurzel aus der Summe der Quadrate seiner Komponenten}*
maintenance *s.* truth maintenance
majority Mehrheit *f*, Majorität *f*
majority-decision gate Majoritäts-Entscheidungsgatter *n*

majority-decision logic Majoritätsentscheidungslogik *f*
majority element Majoritätselement *n*, Majoritätsgatter *n*, Majoritätsschaltung *f*, Majoritätsglied *n*, Mehrheitselement *n*, Mehrheitsgatter *n*, Mehrheitsschaltung *f*, Mehrheitsglied *n*
majority function Majoritätsfunktion *f {Automatentheorie}*
majority game Majoritätsspiel *n*
majority gate Majoritätsgatter *n*
majority logic Mehrheitslogik *f*
majority-minority logic Majoritäts-Minoritäts-Logik *f*
majority network *s.* majority gate
majority system Mehrheitssystem *n*
man Mensch *m*
man-machine communication Mensch-Maschine-Kommunikation *f*
man-machine dialogue Mensch-Maschine-Dialog *m*, Dialog *m* zwischen Mensch und Maschine
man-machine interaction Mensch-Maschine-Kopplung *f*
man-machine interface Mensch-Maschine-Schnittstelle *f*, Mensch-Maschine-Interface *n*, Nahtstelle *f* zwischen Mensch und Maschine
man-machine system Mensch-Maschine-System *n*
management Betriebsführung *f*; Verwaltung *f*
management game Unternehmensspiel *n*
management science Leitungswissenschaft *f*
management system Verwaltungssystem *n*, Leitungssystem *n {Beteiligung des Menschen}*

manifold Mannigfaltigkeit *f* *{Gruppentheorie}*
manipulate/to manipulieren, handhaben
manipulation 1. Manipulierung *f*, Handhabung *f* *{Roboter}*; 2. Datenmanipulation *f* *{Beeinflussung von Daten durch Programme bei unterschiedlichen Verwendungen}*
manipulation robot Handhabungsroboter *m*
manipulator-type robot Handhabungsautomat *m*, Manipulator *m*
many-server system System *n* mit vielen Bedienungsapparaten
many-valued logic mehrwertige Logik *f*
map/to abbilden *{Mathematik}*
map 1. Bild *n*, Abbildung *f*; 2. Landkarte *f*; Karte *f* *{Graphentheorie}*
map inversion Abbildungsinversion *f*
mapping Abbilden *n*; Abbildung *f*
marker model Markierungsmodell *n* *{Netzplantechnik}*
marking Markieren *n*; Kennzeichnung *f*, Kennzeichen *n*, Markierung *f*; Beschriftung *f*
marking function Markierungsfunktion *f* *{Moore-Automat}*
Markov chain Markow-Kette *f*
Markov algorithm Markow-Algorithmus *m*
Markov process Markow-Prozeß *m*
masking Überdeckung *f* *{ein zweiter Reiz wird nicht wirksam, weil sich ein erster Reiz noch in der Verarbeitung befindet}*
mass servicing theory Massenbedienungstheorie *f*
master 1. übergeordnet; 2. Leitgerät *n*; 3. *s.* master computer
master computer Leitcomputer *m*, übergeordneter Computer *m*
master manipulator Leitmanipulator *m*, Leithandhabungsautomat *m*
master system übergeordnetes System *n*
match/to anpassen; zusammenpassen; [passend] zuordnen; vergleichen *{auf Übereinstimmung}*
matching Anpassung *f*, Reihenmatch *n*
matching control modelladaptive Regelung *f*
matching problem Anpassungsproblem *n*, Anschlußproblem *n*
mathematical mathematisch
mathematical decision model mathematisches Entscheidungsmodell *n*
mathematical expectation mathematische Erwartung *f*, Erwartungswert *m*
mathematical language model mathematisches Sprachmodell *n*
mathematical linguistics mathematische Linguistik *f*
mathematical logic mathematische (theoretische, symbolische) Logik *f*, Symbollogik *f*, Regellogik *f*
mathematical modelling mathematische Modellierung *f*
mathematical optimization mathematische Optimierung *f*
mathematical representation mathematische Darstellung *f*
mathematical semantics mathematische Semantik *f*
mathematical simulation mathematische Simulierung (Nachbildung, Modellierung) *f*

mathematical statistics mathematische Statistik f
mathematics Mathematik f
mathematization Mathematisierung f
matrix Matrix f
matrix game Matrixspiel n; Zweipersonennullsummenspiel n
matrix notation Matrixschreibweise f
matrix of partial effect Matrix f der partiellen Wirkung
matrix representation Matrixdarstellung f
maximability Maximierbarkeit f
maximable maximierbar
maximal flow problem Maximalflußproblem n
maximax principle Maximaxprinzip n, Optimalitätsprinzip n von Hurwicz
maximin criterion Maximinkriterium n
maximin principle Maximinprinzip n, Optimalitätsprinzip n nach von Neumann
maximin strategy Maximinstrategie f
maximin theorem Maximintheorem n
maximization Maximierung f
maximization algorithm Maximierungsalgorithmus m
maximization criterion Maximierungskriterium n
maximization problem Maximierungsproblem n
maximize/to maximieren
maximizing function Maximierungsfunktion f
maximum Maximum n, Höchstwert m
maximum chain theorem Maximalkettensatz m {Mengenlehre}
maximum error Maximalfehler m
maximum-flow problem Maximalstromproblem n {Graph}
maximum-gain gain optimization Optimierung f nach maximalem Gewinn
maximum principle Maximalprinzip n, Maximumprinzip n
maximum-value search Maximalwertsuche f
maxterm Elementaralternative f, Vollalternative f, Maxterm m
Mealy automaton Mealyscher Automat m {sequentielle Schaltung}
mean expectation mittlerer Erwartungswert m
mean queue size mittlere Warteschlangenlänge f
meaning Bedeutung f {z.B. eines Signals}
meaningless bedeutungslos
means-end analysis Means-end-Analyse f {eine iterative Analysenmethode, bei der die Abweichung des aktuellen Zustandes vom Zielzustand sukzessive verringert wird}
means-end analysis method mit Differenzverringerung arbeitende Analysenmethode f [zur Problemlösung]
means set Menge f der Mittel, Mittelmenge f
measure of indeterminacy Unbestimmtheitsmaß n
measure of interdependence Maß n der gegenseitigen Abhängigkeit
mechanism Mechanismus m

membrane

membrane Membran *f {dünne Zellgewebeschicht, die den Zellkörper des Neurons umgibt}*
memorization Erinnerungsvermögen *n*, Lernvermögen *n*
memory Gedächtnis *n*
memory-free speicherfrei, gedächtnislos
memory hierarchy Speicherhierarchie *f*
memory of events Ereignisspeicher *m {z.B. bei endlichen Automaten}*
memoryless gedächtnislos, speicherfrei
merge/to mischen *{zwei oder mehr Dateien kombinieren}*, vereinigen, fusionieren; mischsortieren
merger graph Verschmelzungsgraph *m*
merger table Verschmelzungstabelle *f*
merging of variables Variablenverschmelzung *f*
mesh Masche *f {in einem Netzwerk}*
mesh network Maschennetz *n*
meshed vermascht
meshed graphs verflochtene Netzpläne (Graphen) *mpl*
meshed structure vermaschte Struktur *f*
message Nachricht *f*
meta compiler Metacompiler *m*
meta knowledge Wissen *n* über das Wissen
meta-level knowledge Wissen *n* über Wissen in anderen Ebenen
meta logic Metalogik *f*
meta mathematics Metamathematik *f*
meta rule Metaregel *f {anderen Regeln übergeordnete Schlußregel}*

meta-stable metastabil
metagame Metaspiel *n*
metaknowledge Metawissen *n*
metalanguage Metasprache *f*
metasymbol Metasymbol *n*
metatheory Metatheorie *f*
method Methode *f*
method of estimation statistisches Schätzungsverfahren *n*
method of generalized gradients Verfahren *n* der verallgemeinerten Gradienten
method of intercepted hyperplanes Schnittebenenmethode *f {Integraloptimierung}*
method of moments Momentenmethode *f*
method of null potential nodes Nullpotentialknotenmethode *f*
method of steepest descent Methode *f* des steilsten Abstiegs
method of test Prüfmethode *f*
method of Zoutendijk *s.* feasible-directions method
microcellular structure mikrozellulare Struktur *f*
microprocessor semantics Mikroprozessor-Semantik *f*
mid-run explanation Selbsterläuterung *f* eines laufenden Programms *{Erklärung der gerade ablaufenden Arbeits- gänge und der bevorstehenden Absichten des Programms}*; Erklärung *f* mitten im Ablauf *{z.B. beim Anhalten während der Abarbeitung eines Programmes}*
migration Veränderung *f* infolge technischer Weiterentwicklung
minimability Minimierbarkeit *f*
minimable minimierbar

minimal automaton minimaler Automat *m*
minimal disjunctive normal (standard) form minimale disjunktive Normalform *f*
minimax Minimax-
minimax criterion Minimaxkriterium *n*
minimax decision rule Minimax-Entscheidungsregel *f*
minimax identification Minimax-identifikation *f*
minimax principle Minimaxprinzip *n*
minimax search Minimax-Suche *f*
minimax strategy Minimaxstrategie *f*, Vorsichtsstrategie *f*
minimax theorem Minimaxtheorem *n*
minimization Minimierung *f*
minimization algorithm Minimierungsalgorithmus *m*
minimization criterion Minimierungskriterium *n*
minimization method due to Quine Minimierungsverfahren *n* nach Quine
minimization of the set of features Minimalisierung *f* des Merkmalssatzes
minimization problem Minimierungsproblem *n*
minimization technique Minimierungsverfahren *n*
minimize/to 1. minimieren; 2. kürzen *{Schaltfunktion}*
minimizing function Minimierungsfunktion *f*
minimum automaton minimaler Automat *m*
minimum cost of fodder problem
s. diet problem
minimum framework Minimalgerüst *n* *{Netzwerk}*
minimum logic minimale Logik *f*
minimum loss optimization Optimierung *f* nach minimalem Verlust
minimum matrix procedure Matrixminimummethode *f*
minimum principle Minimalprinzip *n*
minimum search method Minimumsuchmethode *f*
minimum tree Minimalbaum *m*
minimum value 1. Mindestwert *m*, Kleinstwert *m*, Minimalwert *m*; 2. Skalenanfangswert *m*, Anfangswert *m*
minterm Minterm *m*, Elementarkonjunktion *f*, Vollkonjunktion *f*
mirrored metagame gespiegeltes Metaspiel *n*
mirroring Spiegeln *n*
mistake Fehler *m*, Fehlverhalten *n* *{Mensch}*
mixed gemischt
mixed graph gemischter Graph *m*
mixed strategy gemischte Strategie *f* *{Spieltheorie}*
mixed-strategy space Raum *m* der gemischten Strategien
mixed system gemischtes System *n*
mixing problem Mischungsproblem *n*
mnemonic diagram Mnemoschema *n*
mnemonics Mnemonik *f*
modality logic Modalitätenlogik *f*
mode 1. Art *f*, Weise *f*, Form *f*; 2. Betriebsart *f*; Modus *m*, Arbeitsweise *f*, Betriebsweise *f*; 3. Schwingungsart *f*, Mode *f*, Schwingungs-

typ *m*; 4. Modus *m*, Modalwert *m*; 5. Wert *m* größter Häufigkeit, Gipfelwert *m* *{Statistik}*
model/to modellieren, [als Modell] nachbilden
model Modell *n*, Muster *n*, Vorlage *f*, Ausführungsform *f*
model-based optimization Optimierung *f* auf der Grundlage von Modellen
model behaviour Modellverhalten *n*
model class Modellklasse *f*
model description Modellbeschreibung *f*
model design Modellentwurf *m*
model driving Modellsteuerung *f* *{eine Top-down-Methode}*
model error Modellfehler *m*, Fehler *m* des Modells
model estimation Modellschätzung *f*
model evaluation Modellbewertung *f*
model experiment Modellversuch *m*
model-extremal system Modellextremalsystem *n*
model hierarchy Modellhierarchie *f*
model mapping Modellabbildung *f*
model method Modellmethode *f*
model object Modellobjekt *n*, Modelloriginal *n*
model of a multistage distribution process Modell *n* eines mehrstufigen Verteilungsprozesses
model of a multistage production process Modell *n* eines mehrstufigen Produktionsprozesses
model of objects to be recognized Modell *n* zu erkennender Objekte
model of reaction Reaktionsmodell *n*

model of the nervous cell Modell *n* der Nervenzelle
model of the visual analyzer Modell *n* des visuellen Analysators
model parameter Modellparameter *m*
model-reference adaptive control adaptive Regelung *f* mit Bezugsmodell, modelladaptive Regelung *f*
model-reference adaptive system modelladaptives System *n*
model-reference control system Steuerungssystem *n* mit Modellvergleich
model-reference system Bezugsmodellsystem *n*, System *n* mit Bezugmodell
model relation Modellrelation *f*
model simplification Modellvereinfachung *f*
model simulation Modellsimulierung *f*
model stability Modellstabilität *f*
model theory Modelltheorie *f*
model updating Modellnachführung *f*
model with variable structure Modell *n* mit variabler Struktur
model[l]ability Modellierbarkeit *f*
model[l]able modellierbar
model[l]ing Modellierung *f*, Modellbildung *f*, Nachbildung *f*
model[l]ing method Modellmethode *f*
model[l]ing of recognition and teaching processes Modellierung *f* der Erkennungs- und Lehrprozesse
model[l]ing of sensory systems Modellierung *f* sensorischer Systeme
model[l]ing of systems of molecular biology Modellierung *f* moleku-

larbiologischer Systeme
model[l]ing of the man-machine system Modellierung f des Mensch-Maschine-Systems
model[l]ing of the memory Modellierung f des Gedächtnisses
model[l]ing of the perception process Modellierung f des Wahrnehmungsprozesses
model[l]ing of thinking Modellierung f des Denkens
model[l]ing strategy Modellierungsstrategie f
modification Abwandlung f, Modifikation f, Änderung f, Umbau m, Systemmodifikation f
module Modul m, Baustein m; Modul m *{kleinste Funktionseinheit im Cortex}*; Modul m *{vertikale Anordnung von Zellen}*
modus ponens Modus ponens m, hypothetischer Syllogismus m, Wahrheitsvorwärtsschluß m *{mathematische Argumentform in deduktiver Logik, Kunstwort}*
modus tollens Modus tollens m, Widerlegungsregel f *{deduktive Logik}*
monadic Boolean operator monadischer Boolescher Operator m
monadic language monadische Sprache f
monadic operation monadische Operation f *{Operation mit genau 3einem Operanden}*
monoid Einselement n, Monoid n *{Mengenlehre}*
monotonic monoton, in gleicher Richtung verlaufend
monotonic Boolean function monotone Boolesche Funktion f
monotonic function monotone Funktion f
monotonic inference (reasoning) monotone Folgerung f, monotones Schließen n *{ein einmal festgelegter Fakt kann im Verlauf der Schlußfolgerung nicht mehr verändert werden}* Monotonie f
Moore automaton Moore-Automat m *{sequentielle Schaltung}*
morphism Morphismus m
move/to bewegen *{z.B. Cursor}*; verschieben *{Information}*; umordnen *{Daten}*
multibranched model mehrzweigiges Modell n
multibranched network mehrfach verzweigtes Netz[werk] n
multifingered hand Hand mit mehreren Fingern, Greifer m mit mehreren Fingern *{Roboter}*
multilevel system Mehrebenensystem n
multiple inheritance mehrfache Vererbung f *{eine Klasse kann von vielen Oberklassen abgeleitet werden}*
multivalued attribute mehrwertiges Attribut n
mutual exclusion gegenseitiger (wechselseitiger) Auschluß m *{Logik}*

N

n-place predicate n-stelliges Prädikat *n*
n-valued logic n-wertige Logik *f*
naive physics naive Physik *f* *{physikalische Erfahrungsgesetze, wie Reibung, Schwerkraft usw.}*
name Identifikator *m*, Bezeichner *m*; Name *m* *{dargestellt als String aus einer vorgegebenen Anzahl von Zeichen}*
NAND circuit (element, gate, network) NAND-Schaltung *f*, NAND-Element *n*, NAND-Glied *n*, NAND-Gatter *n*
NAND operation NAND-Operation *f*, NICHT-UND-Operation *f*, negierte Konjunktion *f*
narrow/to einengen *{z.B. Suchvorgang}*
narrow predicate calculus Prädikatenkalkül *m* erster Stufe
narrow system Schmalspursystem *n* *{eng definiertes Expertensystem}*
native language Muttersprache *f*
natural natürlich
  natural deduction natürlicher Schluß *m*, informelle Schlußfolgerung *f*
  natural intelligence natürliche Intelligenz *f*
  natural language natürliche Sprache *f*
  natural language communication Kommunikation *f* in natürlicher Sprache
  natural-language interface Interface *n* (Schnittstelle *f*) zur natürlichen Sprache *{zum Sprachdialog zwischen Mensch und Computer}*; Interface *n* für natürliche Sprache, natürlichsprachiges Interface *n*
  natural-language processing Verarbeitung *f* natürlicher Sprache
  natural-language translation Übersetzung *f* natürlicher Sprache
  natural-stability limit Eigenstabilitätsgrenze *f*, Grenze *f* der natürlichen Stabilität
  natural system natürliches System *n*
  natural transient stability limit Grenze *f* der natürlichen dynamischen Stabilität *{eines Übertragungssystems}*
near-complete decomposability nahezu vollständige Zerlegbarkeit *f*
nearest neighbour method Nächste-Nachbarn-Methode *f* *{Klassifizierung}*
nearest neighbour rule Nächste-Nachbarn-Regel *f*, NN-Regel *f*
NEARLY ALWAYS FAST IMMER *{unscharfer Quantor}*
NEARLY NEVER FAST NIE *{unscharfer Quantor}*
necessary condition notwendige Bedingung *f* *{Logik}*
negate/to negieren, [eine Aussage] in das Gegenteil verkehren; invertieren
negative entropy negative Entropie *f*, Negentropie *f*
negative example Gegenbeispiel *n* *{künstliche Intelligenz}*
negative-true binary signal Binärsignal *n* in negativer Logik
negative-true logic negative Logik *f*
negator Negator *m*, NICHT-Element *n*, NICHT-Glied *n*, Inverter *m*

negentropy Negentropie *f*, negative Entropie *f*
neighbour Nachbar *m*
neighboured nodes benachbarte Knotenpunkte *mpl*
neighbourhood set Nachbarneuronen *npl* *{Gesamtheit der einem speziellen Neuron unmittelbar benachbarten Neuronen}*
NEITHER-NOR WEDER-NOCH *{Schaltalgebra}*
neocortex Neocortex *m* *{menschlicher Cortex, evolutionär neuester Teil des Cortex}*
nerve cell Nervenzelle *f*
nerve impulse Nervenimpuls *m*
net of neurons Neuronennetz *n*
netware Netware *f* *{Software für neuronale Netzwerke}*
network Netzwerk *n*, Netz *n*, Schaltung *f*; *{s.a.* circuit*}*
network analysis Schaltungsanalyse *f*, Netzanalyse *f*, Netzwerkanalyse *f*, Netzplananalyse *f*
network analyzer Netzanalysator *m*, Netzwerkanalysator *m*
network architecture Netzarchitektur *f*
network control language Netzsteuersprache *f*
network management system Netzverwaltungssystem *n*
network matrix Netzplanmatrix *f*
network model 1. Netzmodell *n*, Schaltungsmodell *n*; 2. Netzplanmodell *n*
network of interconnected computers Computernetzwerk *n*
network parameter Netzparameter *m*
network theory Netzwerktheorie *f*

neural neural, Neuronen-
neural computer neuraler Computer *m*, Neuralcomputer *m*
neural network Neuralnetz *n*; neuronaler Schaltkreis *m*, neuronales Netzwerk *n*
neuristor Neuristor *m* *{Element zur Nachbildung von Neuroneneigenschaften}*
neurobionics Neurobionik *f*
neurocomputer Neurocomputer *m* *{Computer zur Simulierung neuronaler Netzwerke}*
neurocybernetics Neurokybernetik *f*
neurology Neurologie *f*
neuron Neuron *n* *{Nervenzelle oder Verarbeitungselement in einem künstlichen neuronalen Netzwerk}*
neuron-like network neuronenähnliches Netz (Netzwerk) *n*
neuron simulation Neuronensimulation *f*
neutral position Nullstellung *f*, Ruhestellung *f*
neutral state Neutralzustand *m*
neutral zone Unempfindlichkeitsbereich *m*
neutralizing capacitor Entkopplungskondensator *m*, Neutralisationskondensator *m*
NEVER NIE *{Quantor}*
nilpotency Nilpotenz *f*
nilpotent event nilpotentes Ereignis *n*
nine-footway problem Problem *n* der neun Fußwege, Problem *n* der drei Häuser und drei Brunnen, Problem *n* der zänkischen Nachbarn *{Graphentheorie}*
NLI *s.* natural-language interface
NLP *s.* natural-language processing
NN *s.* neuronal network

nodal Knoten betreffend, Knoten- *{s.a.* node*}*
nodal point Knotenpunkt *m* *{z.B.Graph}*
nodal vector Knotenpunktvektor *m*
node 1. Knoten *m*, Knotenpunkt *m* *{z.B. bei einem Netzwerk, einer Baumstruktur oder der Recherche in intelligenten Wissensbasen}*; 2. Knoten *m*, Schwingungsknoten *m*; 3. Stützstelle *f* {Interpolation}
node base Knotenbasis *f {Graph}*
node equation Knotengleichung *f*
node-evaluated graph knotenbewerteter Graph *m*
node evaluation Knotenbewertung *f {Graph}*
node fusion Knotenverschmelzung *f {Graph}*
node input Knoteneingang *m*
node-oriented graph knotenorientierter Netzplan (Graph) *m*
node output Knotenausgang *m*
node potential method Knotenpotentialmethode *f*
node set Knotenmenge *f {Graph}*
node splitting Knotenspaltung *f {Graph}*
node variable Knotenvariable *f {Graph}*
noise Rauschen *n {unzutreffende oder ungenaue Daten in einem Eingabemuster für ein neuronales Netzwerk}*
noise-embedded signal verrauschtes Signal *n*
noise saturation Rauschsättigung *f*
noise suppression Rauschunterdrückung *f*
noiseword unwesentliches Wort *n* [in einem Kontext] *{Wort mit keinem oder nur geringem Informationsgehalt, z. B., der, die, das, von, aus etc.)*
noisy verrauscht, mit Fehlern behaftet
noisy system verrauschtes System *n*
non-algorithmic nichtalgorithmisch
non-algorithmic process nichtalgorithmischer Prozeß *m*
non-ambiguous relation eindeutiger Zusammenhang *m*
non-autonomous system nichtautonomes System *n*
non-branched unverzweigt
non-branched algorithm unverzweigter Algorithmus *m*
non-chronological nichtchronologisch
non-chronological backtracking relevante Rückwärtsverkettung *f {Verkettung zum zweckmäßigsten, jedoch nicht zum letzten Knoten des Suchbaumes}*
non-classical logic nichtklassische Logik *f*
non-classical variational problem nichtklassisches Variationsproblem *n*
non-computable unberechenbar, nicht berechenbar
non-conjunction negierte Konjunktion *f*, NICHT-UND-Operation *f*, NAND-Operation *f*
non-contradictory widerspruchsfrei *{Logik}*
non-contrary interest game Spiel *n* mit nichtentgegengesetzten Interessen
non-convex nichtkonvex *{Optimierung}*
non-cooperative game nichtkooperatives Spiel *n*

non-cooperative game theory nichtkooperative Spieltheorie *f*
non-critical activity nichtkritische Aktivität *f* {*Netzplan*}
non-critical point unkritischer (nichtsingulärer) Punkt *m*
non-decidable unentscheidbar
non-determinated system nichtdeterminiertes System *n*
non-deterministic algorithm nichtdeterministischer Algorithmus *m*
non-deterministic automaton nichtdeterministischer Automat *m*
non-deterministic signal nichtdeterministisches Signal *n*
non-directed edge ungerichtete Kante *f* {*Graph*}
non-directed graph ungerichteter Graph *m*
non-directed tree ungerichteter Baum *m* {*Graph*}
non-directional richtungsunabhängig, nichtgerichtet, ungerichtet
non-discernible event nichtunterscheidbares Ereignis *n*
non-disjunction negierte Disjunktion *f*, NICHT-ODER-Operation *f*, NOR-Operation *f*
non-empty set nichtleere Menge *f*
non-equivalence Antivalenz *f*, ausschließende ODER-Operation *f*; {*s.a.* exclusive OR}
non-equivalence element Antivalenzelement *n*
non-equivalence operation Antivalenzoperation *f*
non-feasible solution unzulässige Lösung *f*
non-feasible strategy unzulässige Strategie *f*
non-functional element nichtfunktionales Element *n*
non-homogeneous inhomogen, nichthomogen
non-homogeneous flow in the transport network inhomogener Fluß *m* im Transportnetz
non-homogeneous transport network problem inhomogenes Transportnetzproblem *n*
non-initial automaton nichtinitialer Automat *m*
non-interacting system entkoppeltes (nichtwechselwirkendes) System *n*
non-interaction Entkopplung *f*
non-interactive wechselwirkungsfrei
non-inverting logic nichtinvertierende Logik *f*
non-linear nichtlinear, von mehr als einer Dimension
non-linear learning nichtlineares Lernen *n*, Lernen *n* in nichtstationärer Umgebung, Lernen *n* unter dem Einfluß von Störgrößen
non-linearity Nichtlinearität *f*
non-Markov process nicht-Markowscher Prozeß *m*
non-monotonic inference nichtmonotones Schließen (Folgern) *n* {*auch als wahr erkannte Werte können wieder verworfen werden*}
non-monotonic logic nichtmonotone Logik *f* {*Schlußfolgerungen können revidiert werden, wenn neue Ergebnisse vorliegen*}
non-monotonic reasoning nichtmonotones Schließen *n*
non-negativity condition Nichtnegativitätsbedingung *f* {*z.B. Optimierung*}

non-optimum estimation   nichtoptimale Schätzung *f*
non-oriented graph   ungerichteter Graph *m*
non-oscillatory instability   monotone Instabilität *f*
non-parametric test   parameterfreier (nichtparametrischer, verteilungsfreier) Test *m*
non-planar graph   nichtplanarer Graph *m*
non-predictive   unvorhersehbar
non-procedural language   nichtprozedurale Sprache *f* *{z.B. Prolog}*
non-random   nichtzufällig
non-sequential system   nichtsequentielles System *n*
non-smooth function   nichtglatte Funktion *f*
non-stationary random process   instationärer Zufallsprozeß *m*
non-strategic game   nichtstrategisches Spiel *n*
non-symmetric   unsymmetrisch, asymmetrisch
non-symmetric logic   unsymmetrische Logik *f*
non-symmetric relation   nichtsymmetrische Relation *f*
non-synthesizable   nichtsynthetisierbar
non-terminal node   nichtterminaler (weiterzerlegbarer) Knoten *m* *{Baumstruktur}*
non-threshold logic   schwellenwertfreie Logik *f*, NTL
non-unique   mehrdeutig
non-unique mapping   mehrdeutige Abbildung *f* *{Korrespondenz}*
non-zero probability   von Null verschiedene Wahrscheinlichkeit *f*
non-zero sum game   Nichtnullsummenspiel *n*
non-zero term   von Null verschiedenes Glied *n*
NOR operation   NOR-Verknüpfung *f*
NOR-function   NOR-Funktion *f*
normalization   Normierung *f*
northwest corner rule   Nordwesteckenregel *f*
NOT   NICHT *n* *{Wahrheitsfunktion}*
NOT AND   NICHT-UND *n*
NOT AND gate   NICHT-UND-Gatter *n*
NOT circuit   NICHT-Schaltung *f*
NOT element   NICHT-Element *n*, NICHT-Glied *n*, Negator *m*, Inverter *m*
NOT function   NICHT-Funktion *f*
NOT gate   Negator *m*, NICHT-Gatter *n*
NOT operation   NICHT-Operation *f*, Negation *f*
NOT OR   NICHT-ODER *n*
NOT OR operation   NICHT-ODER-Operation *f*, NOR-Operation *f*, negierte Disjunktion *f*
notation convention   Notationsvereinbarung *f*
novelty detector   Neuheitsdetektor *m* *{eine adaptive Schaltung, die Änderungen im Tupel erkennt und auf einen von Null verschiedenen Wert geht, wenn das Eingabemuster neu ist}*
nucleus   Nucleus *m*, Kern *m* *{Ansammlung von Nervenzellen in der grauen Gehirnsubstanz}*
null-class law   Gesetz *n* von der Existenz des Nullelements
null graph   Nullgraph *m*
null hypothesis   Nullhypothese *f*

null set leere Menge *f*
number cruncher "Zahlenknakker" *m* {*ein Arithmetikprozessor, oft auch für Großrechner verwendet, die nur zur Arithmetik bei großen Zahlenmengen dienen*}
number field Zahlkörper *m*
number theory Zahlentheorie *f*
numeric mathematics numerische Mathematik *f*
numeric model numerisches Modell *n*

O

object Objekt *n*; {*s.a.* frame}
object description Objektbeschreibung *f*
object generation Objekterzeugung *f*
object language Objektsprache *f*, aufgabenorientierte Sprache *f*
object model Objektmodell *n*
object-oriented programming objektorientiertes Programmieren *n*
object-oriented representation objektorientierte Darstellung *f*
object-oriented system objektorientiertes System *n*
object recognition Objekterkennung *f*
object relation Objektbeziehung *f*
object system Objektsystem *n*
object tree Objektbaum *m*, Themenbaum *m*, Kontextbaum *m*
objective Gesichtspunkt *m*, Ziel *n*
objective condition Zielbedingung *f*
objective coordination Zielkoordination *f*, Preismethode *f*

objective event Zielereignis *n* {*Netzplantechnik*}
objective function vector Zielfunktionsvektor *m*
objective hierarchy Zielhierarchie *f*
objective quantity Zielgröße *f*
objective set Zielmenge *f*
objective value Zielwert *m*
obtain/to erhalten; gewinnen {*Lösung eines Problems*}
obtain an extremum/to einen Extremwert bestimmen (gewinnen)
obtainment time Beschaffungszeit *f* {*Lagerhaltungstheorie*}
occupied place besetzter Platz *m* {*Bedienungstheorie*}
occur/to vorkommen, auftreten {*z.B. in einem logischen Ausdruck*}
occurrence 1. Ereignis *n*; 2. Vorgang *m* {*z.B. in der Netzplantechnik*}
occurrence arrow Vorgangspfeil *m*
occurrence node Vorgangsknoten *m*
occurrence probability Ereigniswahrscheinlichkeit *f*, Eintrittswahrscheinlichkeit *f*, Wahrscheinlichkeit *f* des Auftretens
OFTEN OFT {*unscharfer Quantor*}
olfactory olfaktorisch, Geruchs-
on-the-job training Einschulung *f* am Arbeitsplatz
one-dimensional language eindimensionale Sprache *f* {*Darstellung meist als eindimensionale Zeichenkette*}
one-to-one function Injektion *f*, eineindeutige Funktion *f*
one-to-one homomorphism umkehrbar eindeutiger Homomorphismus *m*

one-to-one relation eineindeutige (umkehrbar eindeutige) Beziehung *f*
ONLY THEN IF NUR DANN, WENN
OOP *s.* object-oriented programming
OOS *s.* object-oriented system
open oriented walk offene gerichtete Pfeilfolge *f*
open queueing system offenes Bedienungssystem (Warteschlangensystem) *n*
open structure offene Struktur *f*
open system offenes System *n*
open walk offene Kantenfolge *f*
open-thermocouple device Thermoelementbruchsicherung *f*
open-thermocouple test Thermoelementbruchkontrolle *f*
operation [logische] Operation *f* *{Ausführung eines einzelnen Befehls}*
operational game Planspiel *n*
operational principle of a computer architecture Betriebsprinzip *n* einer Computerarchitektur
operationalization Wissenskompilierung *f*
operations research Operationsforschung *f*, Unternehmensforschung *f*, Verfahrensforschung *f*, betriebliche Verfahrensforschung *f*
operator Operator *m*
optimal disjunctive normal (standard) form optimale disjunktive Normalform *f*
optimal filter optimales Filter *n*, Optimalfilter *n*
optimal filtering optimale Filterung *f*
optimal filtering problem Optimalfilterproblem *n*

optimal parameter optimaler Parameter *m*, optimale Kenngröße *f*
optimal strategy optimale Strategie *f*, Optimalstrategie *f*
optimal system optimales (optimiertes) System *n*
optimal trajectory optimale Trajektorie *f*
optimal value Optimalwert *m*, Optimum *n*
optimality Optimalität *f*
optimality condition Optimalitätsbedingung *f*
optimality criterion Optimalitätskriterium *n*, Gütekriterium *n*
optimality criterion of Bellman Bellmansches Optimalitätskriterium *n*
optimality criterion of Savage Optimalitätskriterium *n* von Savage, Prinzip *n* des kleinsten Bedauerns, Analyse *f* der entgangenen Gewinne
optimality model Optimalitätsmodell *n*
optimality principle Optimalitätsprinzip *n*
optimalize/to *s.* optimize/to
optimizability Optimierbarkeit *f*
optimizable optimierbar
optimization Optimierung *f*
optimization algorithm Optimierungsalgorithmus *m*
optimization criterion Optimierungskriterium *n*
optimization problem Optimierungsaufgabe *f*, Optimierungsproblem *n*
optimization strategy Optimierungsstrategie *f*
optimization subject to several criteria Optimierung *f* nach mehreren Zielkriterien

optimize/to optimieren
optimizing computer optimierender Computer *m*
optimizing problem Optimierungsproblem *n*
optimizing system optimierendes System *n*, Optimalwertsystem *n*
optimum optimal
optimum Optimum *n*, Optimalwert *m*
optimum behaviour optimales Verhalten *n*, optimaler Verlauf *m*
optimum compromise set optimale Kompromißmenge *f*, Pareto-Menge *f*
optimum condition optimaler Zustand *m*
optimum design optimaler Plan *m*
optimum estimation Optimalwertschätzung *f*, optimale Schätzung *f*
optimum feasible-direction method Methode *f* der optimal brauchbaren Richtung *{Gradientenmethode}*
optimum principle Optimalprinzip *n*
optimum solution optimale Lösung *f*
optimum strategy optimale Strategie *f*, Optimalstrategie *f*
optimum switching function optimale Schaltfunktion *f*
optimum-time system zeitoptimales System *n*
optimum trajectory optimale Trajektorie *f*
OR/to über ODER-Gatter verknüpfen (zusammenschalten)
OR 1. ODER *n*; 2. *s.* operations research
OR circuit ODER-Element *n*, ODER-Glied *n*, ODER-Gatter *n*
OR connective ODER-Verknüpfung *f*

OR element *s.* OR circuit
OR-ELSE ENTWEDER-ODER *n*, ausschließendes (exklusives) ODER *n*
OR function ODER-Funktion *f*
OR gate ODER-Gatter *n*
OR module ODER-Baustein *m*
OR network *s.* OR circuit
OR node ODER-Knoten *m* *{Knoten im Suchbaum, wobei nur einer von vielen erfüllt zu sein braucht}*
OR operation *s.* 1. inclusive OR operation; 2. exclusive OR operation
OR operator ODER-Operator *m*
oral control Bedienung *f* durch Spracheingabe
order/to 1. anordnen *{in einer Ordnung}*; 2. bestellen
order 1. Ordnung *f*, Größenordnung *f*; 2. Ordnung *f*, Reihenfolge *f*; 3. Befehl *m*, Anweisung *f*; *{s.a.* instruction*}*; 4. Bestellung *f*
order of precedence *s.* priority ranking
order relation Anordnungsrelation *f*, Ordnungsrelation *f*
ordered geordnet
ordered pair geordnetes Paar *n* *{Mengenlehre}*
ordered set geordnete Menge *f*
ordered tree geordneter Baum *m*
ordering Ordnen *n*
ordering bias systematische Ordnungsabweichung *f*, Ordnungstendenz *f*
ordering relation *s.* order relation
oriented gerichtet
oriented edge Pfeil *m*
oriented graph gerichteter Graph *m*
oriented path gerichteter Weg *m*

oriented trail gerichteter Weg *m*
oriented walk gerichtete Pfeilfolge *m*
origin Ursprung *m*
original value Ausgangswert *m*
orthogonal 1. orthogonal, rechtwinklig; 2. beziehungslos [zueinander]; 3. linear unabhängig [voneinander]
  orthogonal central design orthogonale zentrale Planung *f* *{Versuchsplanung}*
  orthogonal search orthogonale Suche *f* *{Optimierung}*
orthogonality Orthogonalität *f*
oscillatory behaviour Schwing[ung]sverhalten *n*
output/to ausgeben
output logic Ausgangslogik *f*
output neuron Outputneuron *n*, Ausgabeneuron *n* *{sendet Informationen an den Menschen oder anderswohin}*
output process *s.* queueing output process
output queue Ausgabewarteschlange *f*
outside world Außenwelt *f*
outstar Outstar *m* *{eine neuronale Struktur, die Signale abtastet}*
overall stability Gesamtstabilität *f*
  overall strategy Gesamtstrategie *f*
  overall system Gesamtsystem *n*
overflow system Überlaufsystem *n* *{Bedienungstheorie}*
overflow traffic Überlaufverkehr *m* *{Bedienungstheorie}*
overlay/to überlagern
overlay tree Überlagerungsbaum *m*
overriding input dominierender Eingang *m*
overshadowing Überdeckung *f* *{Einfluß eines bedingten Verhaltens auf ein anderes, wenn z.B. auf eine gegebene Menge Stimuli ein erwarteter Folgezustand eintritt, werden alle anderen als irrelevant angesehen}*

P

packing problem Packungsproblem *n* *{Graphentheorie}*
paint spraying robot Farbsprühroboter *m*, Lacksprühroboter *m*
pair Paar *n*
  pair of characteristics Merkmalspaar *n*
paired graph paarer (bichromatischer) Graph *m*
paper-based information gedruckte (geschriebene) Information *f*
parabolic optimization parabolische (quadratische) Optimierung *f*
paradigmatic relation paradigmatische Relation *f*
parallel parallel
  parallel automaton parallel arbeitender Automat (Computer) *m*
  parallel computer Parallelcomputer *m*, Simultancomputer *m*
  parallel edges parallele Kanten *fpl*, Mehrfachkanten *fpl* *{Graph}*
  parallel interconnection Parallelverknüpfung *f*
  parallel machine *s.* parallel computer
  parallel network Parallelschaltung *f*; Parallelnetz *n* *{Computerarchitektur}*
  parallel processing Parallelverarbeitung *f*, gleichzeitige

Verarbeitung *f*, Nicht-von-Neumann-Ver- arbeitung *f* *{Gegenteil: sequentielle Verarbeitung}*
parallel processing system Parallelverarbeitungssystem *n*
parallel processor system Parallelprozessorsystem *n*
parallel system Parallelsystem *n*
parallel-work conflict Parallelkonflikt *m*, Konflikt *m* bei Parallelarbeit *{z.B. bei gekoppelten Prozessoren}*
parallelism Parallelismus *m*
parallelization of an algorithm Parallelisierung *f* eines Algorithmus
parallelizing strategy Parallelisierungsstrategie *f*
parameter Parameter *m*, Kenngröße *f*, Bestimmungsgröße *f*, Zustandsgröße *f*, Größe *f*; *{s.a. process parameter}*
parameter adapt[at]ion Parameteranpassung *f*, Parameteradaption *f*
parameter-adaptive parameteradaptiv
parameter-adaptive system parameteradaptives System *n*
parameter class Parameterklasse *f*
parameter-dependent parameterabhängig
parameter fluctuation Parameterschwankung *f*, Parameteränderung *f*
parameter-free parameterfrei
parameter-independent parameterunabhängig
parameter-insensitive parameterunempfindlich
parameter space Parameterraum *m*
parameter variation Parameterveränderung *f*, Parametervariation *f*
parameter vector Parametervektor *m*

parametric parametrisch
parametric coding of speech parametrische Sprachcodierung *f*
parent Vorgänger *m* *{z.B. im Suchbaum}*
parent-child relation Vorgänger-Nachfolger-Beziehung *f* *{z.B. von Frame zu Subframe}*
parent node Vorgängerknoten *m*
pareto optimum Pareto-Optimum *n*
pareto set Pareto-Menge *f*, optimale Kompromißmenge *f*
parser Parser *m* *{Computerprogramm zur linguistischen Textanalyse}*; Sprachanalysator *m*
parsing Parsing *n* *{syntaktische Sprachanalyse}*
part 1. Teil *n* *(m)*; Bestandteil *m*; 2. Einzelteil *n*, Teil *n*
partial teilweise; Teil-
partial coalition Teilkoalition *f* *{Spieltheorie}*
partial conjunction Teilverknüpfung *f* *{logische Verknüpfung, die Teil einer größeren Gesamtheit von Verknüpfungen ist}*
partial expression Teilausdruck *m* *{Logik}*
partial set Teilmenge *f*
partially decidable set rekursiv aufzählbare Menge *f*
partially determined automaton partieller (teilweise determinierter) Automat *m*
partially ordered set teilweise geordnete Menge *f*
parts handling machine (robot) Werkstückhandhabungsautomat *m*, Industrieroboter *m* zur Werkstückhandhabung
path 1. Pfad *m* *{z.B. Weg durch einen Graphen}*; 2. Weg *m*, Pfad *m*,

patience 82

Bahn *f* {*z.B. numerische Steuerung*}; 3. Zweig *m* {*Netzwerk*}
path-connected zusammenhängend {*Punktmenge*}
path control Wegsteuerung *f* {*Roboter*}
path of integration Integrationsweg *m*
path robot streckengesteuerter Roboter *m*, Stetigbahnroboter *m*
patience time Geduldzeit *f* {*Bedienungstheorie*}
patient Patient *m*
patient monitoring Patientenüberwachung *f*
patient monitoring system Patientenüberwachungssystem *n*
patient-oriented data patientenorientierte Daten *npl*
pattern Strukturmuster *n*, Muster *n*, statisches Signal *n*, Konfiguration *f*, Form *f* {*eines Signals*}, Objekt *n*, Anordnung *f*, Figur *f*
pattern analysis Musteranalyse *f*
pattern associator Musterassoziationsnetz *n* {*ein Netzwerk, das zwei Muster einander zuordnet oder vergleicht*}
pattern class Objektklasse *f*
pattern classification Objektklassifizierung *f*, Musterklassifizierung *f*
pattern-classifying system objektklassifizierendes System *n*, musterklassifizierendes System *n*
pattern-directed invocation mustergerichtetes Aufrufen *n* {*Aktivieren von Prozeduren durch Vergleich mit gespeicherten Mustern*}
pattern generator Mustergenerator *m*
pattern interpreter Bildauswertegerät *n*, Musterauswertegerät *n*
pattern matching Mustervergleich *m*
pattern of behaviour Verhaltensmuster *n*
pattern positioning Bildpositionierung *f* {*Bilderkennung*}
pattern processing Bildverarbeitung *f*
pattern recognition Mustererkennung *f*, Gestalterkennung *f*, Objekterkennung *f*, Bilderkennung *f*, Zeichenerkennung *f*; Strukturerkennung *f*
pattern recognizer Mustererkennungsgerät *n*, Strukturerkennungsgerät *n*
pattern retrieval Bildwiederauffindung *f*, Musterwiederauffindung *f*
pattern scanner Bildabtaster *m*
pause/to anhalten, warten
paying-off matrix Auszahlungsmatrix *f*
Peirce arrow Peirce-Pfeil *m*
Peirce function Peirce-Funktion *f*, NOR-Funktion *f*, negierte ODER-Funktion *f*, Nicod-Funktion *f*
penalty estimation Strafabschätzung *f*, Penaltyabschätzung *f*
penalty function Straffunktion *f*, Penaltyfunktion *f*
penalty function method Straffunktionsmethode *f*, Penaltyfunktionsmethode *f*
penalty function shifting method Strafverschiebungsmethode *f*, Penalty-shifting-Methode *f*
penalty method Strafmethode *f*, Penaltymethode *f* {*Optimierung*}

penalty principle Strafprinzip *n*,
Penaltyprinzip *n*
perceive/to erfassen *{Muster}*
percept/to wahrnehmen
perceptibility Wahrnehmbarkeit *f*,
Erkennbarkeit *f*
perceptible wahrnehmbar, erkennbar
perception Erfassen *n*, Aufnehmen *n*
{z.B. Bild}; Sinneswahrnehmung *f*;
Wahrnehmung *f {Bildung von Hypothesen über Merkmale der Umgebung oder Suche nach Sensorinformationen, die solche Hypothesen bestätigen oder widerlegen}*
perceptor Wahrnehmungselement *n*,
Perzeptor *m*
perceptron Perzeptron *n {ein schwellwertgesteuertes Sensormodell nach Rosenberg}*
perform/to durchführen, verrichten,
ausführen *{Aufgabe, Tätigkeit}*; leisten
perform a logic operation/to eine
logische Operation ausführen
performability Erfüllbarkeit *f*
{logischer Ausdruck}, Durchführbarkeit *f*
performable erfüllbar, durchführbar
performance aid Ausführungshilfe *f*
performance system Performanzsystem *n {Komplex aus Handlungssystem und Systemumgebung}*
performing coalition Handlungskoalition *f*
performing input Handlungseingang *m {Entscheidungstabelle}*
permissible strategy zulässige Strategie *f*
Petri net Petri-Netz *n*
phenomenon Phänomen *n*; Vorgang *m*

phoneme Phonem *n*
phonetics Phonetik *f*
phrase Phrase *f*
phrase structure grammar Phrasenstrukturgrammatik *f*
phrase structure rule Phrasenstrukturregel *f*
physical machine realer Automat *m*
physical model physikalisches
Modell *n*
physical modelling physikalische
Modellierung *f*
physical simulation system
physikalisches Simulationssystem *n*
pick-and-place robot Handhabungsroboter *m*
pick-and-place unit Aufnahme-
und Plaziereinheit *f*, Handhabungseinheit *f*
pipeline/to im Pipelinesystem verarbeiten
pipeline 1. Pipeline *f*, Rohrleitung *f*;
2. Pipeline *f {Datenverarbeitung}*
pipeline computer Pipelinerechner *m*
pipeline principle Pipelineprinzip *n*
pipeline system Pipelinesystem *n*
{fließbandähnliche Datenverarbeitung in Mehrprozessorsystemen}
pipelining Fließbandbearbeitung *f*,
Pipeline-Verarbeitung *f {Datenverarbeitung}*
pitch axis Nickachse *f {Roboter}*
pivot column Pivot-Spalte *f*
pivot column search Pivot-Spaltensuche *f*
pivot search Pivot-Suche *f*
plain graph schlichter Graph *m*
plan Plan *m {Aktionsfolge zur*

planar 84

Umwandlung einer Anfangssituation in eine Zielsituation}
plan game Planspiel *n*
planar graph planarer (ebener) Graph *m*
plane Ebene *f*
plane graph ebener Graph *m*
plane of support Stützebene *f*
plasticity Anpassungsfähigkeit *f*, Plastizität *f {Fähigkeit einer Gruppe von Neuron, im Laufe der Zeit an eine neue Aufgabe anzupassen und dadurch die Funktion eines beschädigten Teiles eines neuronalen Netzes zu übernehmen}*
plausibility tree Plausibilitätsbaum *m*
plausible plausibel, offensichtlich
plausible inference plausibler Schluß *m {logischer Schluß aus unvollständigen oder ungenauen Vorgaben}*
playback/to wiederholen, abspielen, repetieren *{Band}*; abfahren *{Roboter}*
playback robot Playbackroboter *m*, durch Abfahren und Speichern frei programmierbarer Roboter *m*
player Spieler *m {Spieltheorie}*
playing automaton spielender Automat *m*
plex Plex *m {eine baumähnliche Datenstruktur}*
plex data base Plex-Datenbank *f*
plex structure Plex-Struktur *f*
Poisson process Poissonscher Prozeß *m*
Poisson stream Poissonscher Strom *m {Bedienungstheorie}*
Poisson traffic Poisson-Verkehr *m*
polar coordinate success pattern Erfolgsspinne *f {Entscheidungstheorie}*
pole balancing problem Balanceproblem *n*
polyautomaton Polyautomat *m*
polyflop Schaltkreis *m* mit mehr als zwei möglichen Zuständen; *{s.a.* multistate circuit*}*
polymorphic polymorph
polyoptimization Polyoptimierung *f {Optimierung von Systemen mit mehreren einander widersprechenden Zielfunktionen}*
polyoptimum Polyoptimum *n*
polyvalence Mehrwertigkeit *f*
position-finding probability Ortungswahrscheinlichkeit *f {Suchtheorie}*
position game Positionsspiel *n*
possible event mögliches Ereignis *n {Wahrscheinlichkeitsrechnung}*
post-event Nachereignis *n {Netzplantechnik}*
postcedent Hinterglied *n {Logik}*
postcondition Nachbedingung *f*
postsection Postsektion *f {Logik}*
potency Mächtigkeit *f {z.B. Menge}*
potential problem Potentialproblem *n {Netzwerk}*
pragmatics Pragmatik *f {Beziehung zwischen Zeichen und Strings zu ihrer Interpretation und Verwendung}*
pre-event Vorereignis *n {Netzplantechnik}*
pre-optimization Voroptimierung *f*
preceding activity vorangehende Aktivität *f {Netzplantechnik}*
precompact game präkompaktes (vollbeschränktes) Spiel *n*
precompact space präkompakter (vollbeschränkter) Raum *m*

precomplete class of functions prävollständige Funktionenklasse f
precompute/to vorausberechnen
precondition Vorbedingung f
predecessor Vorgänger m
predecessor state Vorgängerzustand m
predecision Vorentscheidung f
predetermination Vorausberechnung f, Vorausbestimmung f
predetermine/to vorbestimmen, vorausberechnen, vorgeben *{z.B. Sollwert}*
predetermined vorbestimmt
predetermined state vorgegebener Zustand m
predicate Prädikat n *{Logik, Teil einer Aussage}*
predicate calculus Prädikatenkalkül n; *{s.a. narrow predicate calculus}*
predicate logic Prädikatenlogik f *{auf die Verwendung von Variablen und Variablenfunktionen modifizierte Aussagenlogik}*
predicate-logic functor prädikatenlogischer Funktor (Junktor) m
predicate-logic language prädikatenlogische Sprache f
predicate transformer Prädikatenumformer m
predicate variable Prädikatenvariable f
predicativity Prädikativität f
predict/to vorhersagen, voraussagen, vorausberechnen, prädizieren
predictable vorhersagbar
predicted value vorhergesagter (vorausberechneter) Wert m
predicting filter vorhersagendes Filter n, Prädiktionsfilter n, Vorhaltfilter n
prediction Prädiktion f, Vorhersage f, Voraussage f, Prognose f *{z.B. prospektives System}*; Vorhalt m
prediction algorithm Prädiktionsalgorithmus m, Vorhersagealgorithmus m
prediction method Vorhersagemethode f, Prognosemethode f
prediction problem Vorhersageproblem n
prediction theory of random processes Vorhersagetheorie f für Zufallsprozesse
predictive vorhersehbar
predictive coding prädiktive (vorhersagende) Codierung f
predictive method *s.* prediction method
predictive model Vorhersagemodell n
predictive strategy Vorhersagestrategie f
predictive value Prognosewert m, Vorhersagewert m
predictor Prädiktor m *{Logik}*; Prädiktor m, Vorhersageeinrichtung f
preemptive loss queueing system Bedienungssystem n mit absoluter Priorität mit Verlust
preemptive queueing system Bedienungssystem n mit absoluter (unterbrechender) Priorität
preemptive repeat-identical queueing system Bedienungssystem n mit absoluter Priorität mit identischem Neubeginn
resume queueing system Bedienungssystem n mit absoluter Priorität mit Fortsetzung

preference relation Vorzugsrelation *f*
prefix notation Präfixnotation *f* *{bei der Programmierung in der Programmsprache LISP verwendete Listendarstellung}*
prematched system vorangepaßtes System *n*
premise Vordersatz *m {Logik}*, Voraussetzung *f*, Prämisse *f {Aussage als Ausgangsinformation für eine logische Schlußfolgerung}*
prenex form pränexer Ausdruck *m*
prenex normal form pränexe Normalform *f*
prenode Vorknoten *m*
preordering Präordnung *f*, Quasiordnung *f*
prescriptive logic präskriptive Logik *f*
prescriptive paradigm Rezeptparadigma *n*
presection Präsektion *f {Logik}*
presence Präsenz *f*, Gegenwart *f*, Vorhandensein *n*, Existenz *f*
presentation Darstellung *f*
prevent/to verhindern *{z.B. Feuern eines Neurons}*
primal dual algorithm Primal-Dual-Algorithmus *m*, Stepping-stone-Methode *f {lineare Optimierung}*
primal problem Primalproblem *n*, Grundproblem *n {Simplexmethode}*
primary primär, hauptsächlich
primary description Primärbeschreibung *f*, primäre Objektbeschreibung *f*
primary plane Hauptebene *f*
prime attribute Hauptattribut *n*
prime conjunction Primkonjunktion *f*
prime implication Primimplikation *f*
primitive Primitiv *n*
primitive recursion primitive Rekursion *f*
primitive recursive function primitiv-rekursive Funktion *f*
primitive recursive predicate primitiv-rekursives Prädikat *n*
primitive recursive relation primitiv-rekursive Relation *f*
principle of contradiction Prinzip *n* (Satz *m*) vom ausgeschlossenen Widerspruch *{Aussagenlogik}*
principle of excluded middle third Prinzip *n* (Satz *m*) vom ausgeschlossenen Dritten *{Aussagenlogik}*
principle of extensionality Extensionalitätsprinzip *n {Aussagenlogik}*
principle of indirect conclusion Prinzip *n* des indirekten Schließens
principle of inductive conclusion Prinzip *n* des induktiven Schließens
principle of least squares Prinzip *n* der kleinsten Quadrate, Maximum-Likelihood-Prinzip *n*
principle of locality Lokalitätsprinzip *n*
principle of operation Arbeitsweise *f*
principle of realizability of result Prinzip *n* der Zielrealisierbarkeit
principle of similarity Ähnlichkeitsprinzip *n*
principle of superposition Superpositionsprinzip *n*, Satz *m* der Überlagerung
principles of systems analysis Prinzipien *npl* der Systemanalyse

priority Priorität f, Vorrang m
priority arbitration Prioritätsentscheidung f
priority arbitration logic Prioritätsentscheidungslogik f
priority circuit Prioritätskreis m
priority coder Prioritätscodierer m
priority decision Prioritätsentscheidung f, Vorrangentscheidung f
priority level Prioritätsgrad m, Wertigkeit f, Prioritätsebene f, Vorrangebene f
priority logic Prioritätslogik f
priority queue Bedienungssystem n (Warteschlange f) mit Prioritäten
priority ranking Vorrangordnung f, Rangordnung f
priority sequence Prioritätsfolge m
priority serving Abfertigung f mit Prioritäten {Bedienungstherorie}
probabilistic probabilistisch
probabilistic automaton probabilistischer Automat m, Wahrscheinlichkeitsautomat m
probabilistic logic Wahrscheinlichkeitslogik f, probabilistische Logik f
probabilistic machine Wahrscheinlichkeitsmaschine f
probabilistic model probabilistisches (wahrscheinlichkeitstheoretisches) Modell n
probabilistic process Wahrscheinlichkeitsprozeß m
probabilistic relationship Wahrscheinlichkeitsbeziehung f
probabilistic system probabilistisches System n
probabilistic traffic analysis wahrscheinlichkeitstheoretische Verkehrsanalyse f

probabilistic traffic model wahrscheinlichkeitstheoretisches Verkehrsmodell n
probability Wahrscheinlichkeit f
probability array Wahrscheinlichkeitsfeld n
probability calculation Wahrscheinlichkeitsrechnung f
probability comparison Wahrscheinlichkeitsvergleich m
probability curve Wahrscheinlichkeitskurve f {Fehlerkurve}
probability density Wahrscheinlichkeitsdichte f
probability logic Wahrscheinlichkeitslogik f {Boolesche Algebra}
probability model Wahrscheinlichkeitsmodell n
probability of certainty Wahrscheinlichkeit f des sicheren Auftretens [eines Ereignisses]
probability of waiting Wartewahrscheinlichkeit f
probability-restriction model Modell n mit Wahrscheinlichkeitseinschränkung
probability space Wahrscheinlichkeitsraum m
probability theory Wahrscheinlichkeitstheorie f
probability threshold Wahrscheinlichkeitsschwelle f
probability use in pattern recognition Anwendung f der Wahrscheinlichkeitsrechnung bei der Mustererkennung
problem Problem n, Aufgabe f, Aufgabenstellung f
problem analysis Problemanalyse f
problem analyst Problemanalytiker m

problem approach Problemlösung *f*
problem-characteristic problemcharakteristisch
problem class Problemklasse *f*
problem correction Problemkorrektur *f*
problem definition Problemdefinition *f*
problem definition language Problemdefinitionssprache *f*
problem-dependent problemabhängig
problem-describing language problembeschreibende Sprache *f*
problem description Problembeschreibung *f*
problem determination Problembestimmung *f*
problem diagnosis Problemdiagnose *f*
problem formulation Problemstellung *f*, Aufgabenstellung *f*
problem-independent problemunabhängig, aufgabenunabhängig
problem input Problemeingabe *f*
problem of random walk Problem *n* des Handelsreisenden
problem of shortest way Problem *n* des kürzesten Weges
problem of stability Stabilitätsproblem *n*
problem of the cheapest telephone network Problem *n* des billigsten Telefonnetzes
problem of the distribution of deliveries Problem *n* der Verteilung von Lieferungen
problem of the seven bridges of Koenigsberg Königsberger Brückenproblem *n*

problem of travelling salesman Rundreiseproblem *n*, Rundfahrtproblem *n*
problem-oriented problemorientiert, problembezogen
problem-oriented instruction set problemorientierter Instruktionsvorrat (Befehlvorrat) *m*
problem-oriented language problemorientierte Sprache (Programmiersprache) *f*
problem-oriented notation problemorientierte (problemnahe) Notation *f*
problem-oriented software problemorientierte Software *f*
problem reduction Problemreduktion *f {Aufteilung von Problemen in leichter lösbare Teilprobleme}*
problem reformulation Problemumformulierung *f*
problem situation Problemsituation *f*
problem solution Problemlösung *f*
problem solver Problemlöser *m*
problem solving method Problemlösungsmethode *f*
problem space Problemraum *m*; Problembereich *m*, Suchraum *m*
problem state Problemstatus *m*, Problemzustand *m*
problem time Problemzeit *f {Zeitintervall im simulierten System}*
problem variable Problemvariable *f*, Problemgröße *f*
problem with fixed instants of time Problem *n* mit festen Zeitpunkten
problem with mobile ends Problem *n* mit beweglichen Enden

procedural prozedural
procedural knowledge representation prozedurale Wissensdarstellung *f {Wissensdarstellung mittels kleiner Programme, die bereits Wissen über bestimmte Dinge oder Verrichtungen haben}*
procedural language prozedurale Sprache *f*
procedural model Verfahrensmodell *n*
procedural representation aktionsorientierte Darstellung *f*; Verfahrensdarstellung *f*
proceduralization Wissenskompilierung *f*
procedure Prozedur *f {ein Unterprogramm}*
procedure-oriented language verfahrensorientierte Sprache (Programmiersprache) *f*
procedure-oriented programming system verfahrensorientiertes Programmiersystem *n*
procedure-oriented software package verfahrensorientiertes Programmpaket *n*
process Prozeß *m*; Vorgang *m {s.a. activity}*
process analysis Prozeßanalyse *f {s.a. activity analysis}*
process analyzer Prozeßanalysenmeßeinrichtung *f*, Betriebsanalysenmeßeinrichtung *f*
process complexity Prozeßkomplexität *f*
process-dependent prozeßabhängig
process description Prozeßbeschreibung *f*
process event Prozeßereignis *n*
process-independent prozeßunabhängig
process information Prozeßinformation *f*
process model Prozeßmodell *n*
process model parameter Prozeßmodellparameter *m*
process parameter Prozeßparameter *m*
process parameter estimation Prozeßparameterschätzung *f*
process recognition Prozeßerkennung *f*
process safeguarding Prozeßsicherung *f*
process simulation Prozeßsimulierung *f*
process stabilization Prozeßstabilisierung *f*
process state Prozeßzustand *m*
process state description Prozeßzustandsbeschreibung *f*
process synthesis Prozeßsynthese *f*
processing element Verarbeitungselement *n {im allgemeinen ein Neuron; es hat eine Anzahl Inputs und nur einen einzigen Output}*
processing node Verarbeitungsknoten *m {Informationsübertragung}*
processing section Verarbeitungsabschnitt *m*
production 1. Produktion *f*, Fertigung *f*; 2. Produktion *f {WENN-DANN-Regel in Expertensystemen}*
production rule Produktionsregel *f {s.a. implication}*
production system Produktionssystem *n*, regelbasiertes System *n*
professional workstation intelligenter (computergestützter) Arbeitsplatz *m*
profit Gewinn *m*

profit function Nutzenfunktion *f*, Gewinnfunktion *f*, Auszahlungsfunktion *f*
profit matrix Nutzenmatrix *f*, Gewinnmatrix *f*, Auszahlungsmatrix *f*
prognostic method Prognosemethode *f*, Prognosesystem *n*, Vorhersagesystem *n*
program Programm *n*
program-definable key programmierbare Taste *f*
programmable programmierbar
programmable array logic Logik *f* der programmierbaren Felder, programmierbare Array-Logik *f*
programmable logic array programmierbares logisches Feld *n*
programmable logic development system programmierbares logisches Entwicklungssystem *n*
programmed programmiert
programmed logic programmierte Logik *f*
programmed logic array programmiertes Logikfeld *n*
programmed textbook programmiertes Lehrbuch *n*
programming Programmierung *f*, Programm- {*s.a.* program, optimization}
programming environment Programmierumgebung *f* {*z.B. Sprache, Schnittstelle, Editor u.a.m.*}
programming language Programmiersprache *f*
programming model Programmmodell *n*
programming module Programmiermodul *m*
programming program programmierendes Programm *n*

programming tool Programmierhilfe *f*, Programmierwerkzeug *n*
programming variable Programmiervariable *f*
progress/to ablaufen {*z.B. Vorgang oder Programm*}
progress Ablauf *m*, Fortschritt *m*
progressive reduction progressive (voranschreitende) Reduktion *f* {*Logik*}
projection Projektion *f* {*in eine andere Dimension*}
pronounced extremum ausgeprägtes Extremum *n*
pronounced maximum ausgeprägtes Maximum *n*
pronounced minimum ausgeprägtes Minimum *n*
pronounced optimum ausgeprägtes Optimum *n*
proof 1. Beweis *m*; Nachweis *m*; 2. Schutzmaßnahme *f* {*z.B. gegen unbefugte Benutzung*}
proof by induction induktiver Beweis *m*
proof by recursion Rekursionsbeweis *m*, Beweis *m* durch Rekursion
proof theory Beweistheorie *f*
proof tree Beweisbaum *m*
proofing 1. Beweisführung *f*, 2. Schützen *n* {*z.B. von Daten gegen Unbefugte*}
proper richtig; eigentlich
proper function Eigenfunktion *f*
proper subgroup eigentliche (nichttriviale) Untergruppe *f* {*Gruppentheorie*}
proper subset echte Teilmenge *f*
property Eigenschaft *f*
property list Eigenschaftsliste *f* {*Wissensdarstellung mittels*

Beschreibung des Zustandes eines Systems durch Beschreibung der Objekte des Systems}
proportion Verhältnis *n*, Proportion *f*
proportionality Proportionalität *f*
proportionality range Proportionalitätsbereich *m*
proposition Aussage *f*, logische Aussage (Behauptung) *f*; [logischer] Ausdruck *m {kann wahr oder falsch sein}*
proposition variable Aussagenvariable *f*
propositional calculus Aussagenkalkül *m*
propositional connection Aussagenverknüpfung *f*
propositional connective Junktor *m*
propositional equivalence Bisubjunktion *f*
propositional form Aussageform *f {Prädikatenlogik}*
propositional function Aussagenfunktion *f*
propositional logic Aussagenlogik *f*, Wahrheitswertelogik *f {eine elementare Logik, die mit Hilfe von Argumenten den Wahrheitswert einer aus bekannten Aussagen gewonnenen neuen Aussage herleitet}*
propositional-logic expression aussagenlogischer Ausdruck *m*
propositional-logic functor aussagenlogischer Funktor (Junktor) *m*, aussagenlogische Konstante *f*
propositional-logic identity aussagenlogische Identität *f*
propositional variable Aussagenvariable *f*
prospective automaton prospektiver Automat *m*
prospective system prospektives System *n*, Pro-System *n*
prospector Schürfer *m*, Prospektor *m*
prototype Prototyp *m {Anfangssystem als Ausgangsbasis für die Entwicklung weiterer Systeme}*
prove/to beweisen *{Hypothese}*; zeigen
prove Beweis *m {Hypothese}*
proving Beweisen *n*
proximum Proximum *n*, Bestapproximation *f*
pruning Stutzen *n*, Beschneiden *n {Zweige eines Entscheidungsbaumes}*
pseudoadaptive resonance pseudoadaptive Resonanz *f {tritt auf, wenn die beiden Schichten eines bidirektionalen assoziativen Speichers in Gleichgewicht geraten}*
pseudoaddition Pseudoaddition *f {Boolesche Algebra}*
pseudoconvex optimization pseudokonvexe Optimierung *f*
pseudoreduction Pseudoreduktion *f {Methode zur Aufteilung von Problemlösungen auf einzelne Plansegmente für Teilziele des zu lösenden Problems}*
pseudotimes Pseudomal *n {Boolesche Algebra}*
pseudovariable Pseudovariable *f*, Pseudoveränderliche *f*
psychological model psychologisches Modell *n*
pursue/to verfolgen *{Verfolgungsspiel}*
pursuer Verfolger *m {Verfolgungsspiel}*
pursuit game Verfolgungsspiel *n*

pyramidal cell Pyramidenzelle *f*
{hauptsächliche Zellform der Nervenzellen im zerebralen Cortex}

Q

quantification Quantifikation *f {Logik}*
quantification theory Prädikatenlogik *f*
quantified variable quantifizierte Variable *f {Prädikatenlogik}*
quantifier Quantor *m*, Quantifikator *m*, logischer Operator *m*, logisches Zeichen *n*
quantify/to quantifizieren *{Logik}*
quantized system quantisiertes System *n*
quasi-analog[ue] model quasianaloges Modell *n*
quasi-linear automaton quasilinearer Automat *m*
quasi-optimum quasioptimal
quasi-stability Quasistabilität *f*
quasi-stationary quasistationär
quaternary logic vierwertige Logik *f*
query Suchfrage *f*, Abfrage *f {bei Informationssystemen}*
query language Abfragesprache *f*
question Frage *f*, Anfrage *f*
question-answering system Fragenbeantwortungssystem *n*, antwortendes System *n*
queue/to Warteschlange bilden, als Warteschlange anordnen
queue Warteschlange *f {Bedienungstheorie}*
queue length Warteschlangenlänge *f*
queueing network Bedienungsnetz *n*, Bediengraph *m*
queueing output process Bedienungs-Ausgangsprozeß *m*
queueing problem Warteschlangenproblem *n*
queueing system Bedienungssystem *n*, Warteschlangensystem *n*
queueing system with limited latency time Bedienungssystem *n* mit begrenzter Wartezeit
queueing system with limited residence period Bedienungssystem *n* mit begrenzter Verweildauer
queueing system with time limits Bedienungssystem *n* mit Zeitbegrenzungen
queueing theory Massenbedienungstheorie *f*, Bedienungstheorie *f*, Warteschlangentheorie *f*

R

race Wettfahrt *f*, Wettrennen *n*
racing Wettrennen *n*, Wettlauf *m {bei indifferenten Ansprechbedingungen}*
ramification Verzweigung *f {Graph}*
ramification point Verzweigungspunkt *m*
random 1. zufällig, regellos, willkürlich, [statistisch] verteilt; 2. ungerichtet; 3. wahlfrei *{Zugriff}*
random choice zufällige Auswahl *f*
random coding Zufallscodierung *f*
random event zufälliges Ereignis *n*, Zufallsereignis *n*
random function zufällige (regellose) Funktion *f*, Zufallsfunktion *f*
random logic Zufallslogik *f*

random model statistisches Modell *n*
random parameter system System *n* mit zufällig schwankenden Parametern *{mindestens ein Parameter}*
random-path problem Zufallswegproblem *n*
random quantity zufällige Größe *f*, Zufallsgröße *f*
random search zufällige Suche *f*
random search method Methode *f* der zufälligen Suche, Zufallssuche *f*
random variable zufällige (regellose) Variable (Größe) *f*, Zufallsvariable *f*, Zufallsgröße *f*
random vector Zufallsvektor *m*, mehrdimensionale Zufallsgröße *f*
random walk Zufallsbewegung *f*, Irrfahrt *f*
random-walk method Monte-Carlo-Methode *f*
randomization Randomisierung *f*, Erzeugung *f* von Regellosigkeit
randomize/to randomisieren, regellos anordnen, wahllos verteilen, regellos auswählen
randomness Zufälligkeit *f*, Zufall *m*
range 1. Bereich *m*; 2. Einsatzbereich *m*; 3. Wertereihe *f*; Wertebereich *m*; 4. Nachbereich *m* *{Relation}*;
range of data Wertebereich *m*
range of definition Definitionsbereich *m*
range of use Anwendungsbereich *m*
rank Rang *m*; Datenebene *f*, Rangfolge *f*
ranked hierarchisch geordnet
rating system Bewertungssystem *n*
rating variable Bewertungsvariable *f*, Bewertungsgröße *f*

reading automaton according to correlation principle Leseautomat *m* nach dem Korrelationsprinzip
real 1. echt, wirklich, wahr; 2. wirksam, effektiv; 3. reell
real-process model[l]ing Modellierung *f* realer Prozesse
real situation Realfall *m*
real-world problem reales (echtes, wirkliches) Problem *n*
reason maintenance Erhaltung *f* des Folgerungswegs *{in Wissensbasen bei Hinzukommen neuer Information}*
reasonableness check Plausibilitätskontrolle *f*
reasoning Argumentation *f*, Beweisführung *f*, Schlußfolgerung *f*; Schlußfolgern *n* [auf der Basis von Erfahrungen]
receiving neuron Empfangsneuron *n*
receptor Rezeptor *m*
reciprocity principle Prinzip *n* der Umkehrbarkeit, Umkehrbarkeitsprinzip *n*
recognition Erkennung *f* *{z.B. Mustererkennung}*
recognition memory Erkennungsspeicher *m*
recognition of visual objects Erkennung *f* visueller Objekte
recognition system Erkennungssystem *n* *{z.B. zur Mustererkennung}*
recognizability Erkennbarkeit *f*
recognizable erkennbar
recognize/to erkennen
recognize-act cycle Erkennen-Handeln-Zyklus *m* *{künstliche Intelligenz}*
recognizer Recogniser *m*, Erkennungsgerät *n*

recognizing grammar erkennende Grammatik *f*
reconfigurable rekonfigurierbar, umkonfigurierbar
reconstructable rekonstruierbar *{Systemzustand}*
record 1. Aufzeichnung *f*, Aufnahme *f*; 2. gespeicherte Daten *npl* *{Informationen}*, Datensatz *m*; 3. Protokoll *n*
record playback robot Playback-Roboter *m*, Roboter *m* mit [in einer Lernphase bei Bewegungsführung durch den Bediener] fest gespeichertem Arbeitsablauf
recovery theory Erneuerungstheorie *f*
rectilinear geradlinig
rectilinear-Cartesian (rectilinear-coordinate) robot nach rechtwinkligen Koordinaten arbeitender Roboter *m*
rectilinear coordinates geradlinige Koordinaten *fpl*
recurrence Rekurrenz *f*
recurrent wiederkehrend
recursion Rekursion *f*
recursive rekursiv
  recursive function rekursive Funktion *f*
  recursive game rekursives Spiel *n*
  recursive method Rekursionsmethode *f*
  recursive operation rekursive Operation *f*
  recursive predicate rekursives Prädikat *n*
  recursive procedure rekursive Prozedur *f*
recursively enumerable predicate rekursiv abzählbares Prädikat *n*
recursively enumerable relation rekursiv abzählbare Relation *f*
recursively enumerable set rekursiv abzählbare Menge *f*
reduced automaton reduzierter Automat *m*
reduced grammar reduzierte Grammatik *f*
reductive conclusion Reduktionsschluß *m*, reduktiver Schluß *m*
redundancy Redundanz *f*, Weitschweifigkeit *f*
redundancy optimization Redundanzoptimierung *f*, Optimierung *f* der Redundanz
redundant-axis roboter Roboter *m* mit redundanten Freiheitsgraden
redundant system redundantes System *n*
reference model Bezugsmodell *n*, Vergleichsmodell *n*
reference object Bezugsobjekt *n*
reflexive reflexiv
  reflexive game reflexives Spiel *n*
  reflexive relation reflexive Relation *f*
reflexivity Reflexivität *f*
region of convergence Konvergenzbereich *m* *{Mathematik}*
region of stability Stabilitätsbereich *m*, stabiler Bereich *m*
regression Regression *f* *{Statistik}*
regressive reduction regressive (zurückschreitende) Reduktion *f* *{Logik}*
regular approximation of a function gleichmäßige Approximation *f* einer Funktion
rehearsal Einüben *n*, Festigen *n* *{Gelerntes durch Praxis und Wiederholung}*
reinforcement Verstärkung *f*

reinforcer verstärkender Reiz *m* *{sucht das Verhalten zu verstärken}*
relate/to beziehen *{z.B. eine Größe auf eine Bezugsgröße}*
relation 1. Relation *f*, Beziehung *f*; 2. *s.* relational structure
relation table Beziehungstabelle *f*
relational operator Vergleichsoperator *m*
relational query Vergleichsabfrage *f*
relational structure relationale Struktur *f* *{eine abstrakte Struktur, mit der einzelne Informationen einander zugeordnet werden}*
relationship Relation *f*
relationship of derivability Ableitbarkeitsrelation *f*
relationship of identity Identitätsrelation *f*
relationship of order Anordnungsrelation *f*
relative importance *s.* weight
relevance tree method Zielbaummethode *f*
relevant backtracking relevantes Zurückverfolgen *n* *{bis zu einem Punkt, der zum Zurückverfolgen am meisten relevant ist}*; relevante Rückwärtsverkettung *f* *{zum zweckmäßigsten Knoten, nicht unbedingt zum letzten Knoten des Suchbaumes}*
reliability calculation Zuverlässigkeitsberechnung *f*
reliability of recognition Erkennungszuverlässigkeit *f*
relocatability Verschiebbarkeit *f*, Verschieblichkeit *f*
relocatable relokatierbar, verschiebbar
relocatable expression verschiebbarer Ausdruck *m*

relocatable term verschieblicher Term *m*
remaining discrete logic diskrete Restlogik *f*
remark Kommentar *m*, Anmerkung *f*, Bemerkung *f*
renewal theorem Erneuerungstheorem *n*
renewal theory Erneuerungstheorie *f*, Ersatztheorie *f*
repeatability Wiederholbarkeit *f*, Reproduzierbarkeit *f*
repetitive wiederholbar
replaceability theorem Ersetzbarkeitstheorem *n* *{Logik}*
replacement theory Ersatztheorie *f*, Erneuerungstheorie *f*
replication Replikation *f*, Gegenimplikation *f* *{Logik}*
represent pictorially/to bildlich darstellen
representation Darstellung *f*
representation language Darstellungssprache *f*
representation of knowledge Wissensdarstellung *f*
representation problem Repräsentationsproblem *n* *{Graphentheorie}*
representation theory Darstellungstheorie *f*
requirement Forderung *f* *{Bedienungstheorie}*
resolution Auflösung *f* *{z.B. Konfliktauflösung}*
resonant state Resonanzzustand *m* *{Resonanz zwischen ankommender Information und Rückkopplungserwartungen}*
resource Ressource *f*
resource allocation Ressourcenverteilung *f*

resource-distribution analysis Ressourcenverteilungsanalyse *f*
respond/to ansprechen [auf]
response Verhalten *n* [nach einem Stimulus]
response of pattern recognition system Antwort *f* des Erkennungssystems
response time Ansprechzeit *f*
rest time Pausenzeit *f* *{Bedienungstheorie}*
resting potential Ruhepotential *n*
restrict/to beschränken
restriction Einschränkung *f*
reteach/to neu lehren *{den lernfähigen Roboter}*
retrieve/to 1. wieder[auf]finden, zurückholen; 2. wiedergewinnen, heraussuchen *{gespeicherte Informationen}*; 3. Liste *f* [mit herausgesuchten Informationen] erstellen; 4. abfragen *{Datenbank}*
retrospect Rückblick *m* *{retrospektives System}*
retrospective automaton retrospektiver Automat *m*
retrospective search retrospektive Suche *f*
retrospective system retrospektives System *n*, Retro-System *n*
revertible model umkehrbares Modell *n*
rheolinear system rheolineares System *n*, System *n* mit zeitlich veränderlichen Parametern
right 1. rechts; 2. recht, richtig, zutreffend; ordentlich, gut, geeignet
right-hand side Aktionsteil *m* [einer Regel]
rigid-body kinematics Kinematik *f* des starren Körpers *{Robotik}*

risk case Risikosituation *f*
risk situation Risikosituation *f*
robotic artificial intelligence künstliche Roboterintelligenz *f*
robotics Robotik *f*, Robotertechnik *f*
roll/to rollen *{Roboterbewegung}*
root Wurzel *f*, Nullstelle *f* *{Funktion}*
root node Wurzelknoten (Anfangsknoten) *m* einer Baumstruktur
round-robin model Reigenmodell *n* *{Bedienungstheorie}*
round-robin search Rundumsuche *f*
route/to 1. [um]leiten *{Datenfluß}*; 2. routen, einen Weg verfolgen (suchen), auf dem Signal laufen
route Weg *m*, Route *f*
router Wegverfolger *m*
router code Wegverfolgungscode *m*
routing 1. Wegeermittlung *f*; 2. Leitweg *m*; 3. Fertigungsplanung *f*, Fertigungsvorbereitung *f*; 4. Bahnführung *f*
RSH *s.* right-hand side
rule Regel *f*, Schlußregel *f* *{durch Feuern gestartetes Unterprogramm einer Regel, bestehend aus LHS und RHS}*
rule base Regelbasis *f*
rule-based system regelbasiertes System *n*, Produktionssystem *n*
rule extraction Regelextraktion *f* *{Gewinnung des gewünschten Outputs durch Verallgemeinerung}*
rule generation Regelerzeugung *f*
rule interpreter Regelinterpreter *m* *{Steuerstruktur für die Implikation}*
rule of inference Schlußregel *f* *{Logik}*
rule of thumb Faustregel *f*

rule-oriented regelorientiert
run/to 1. laufen, arbeiten, in Betrieb sein; 2. funktionieren; 3. durchlaufen lassen, abarbeiten *{Programm}*
running mode laufender Betrieb *m* *{das Netzwerk hat gelernt, die Gewichte bleiben konstant und dienen dazu, Entscheidungen zu treffen}*
runtime system Laufzeitsystem *n*, Mantel *m* eines Expertensystems

S

S expression S-Ausdruck *m* *{ein symbolischer Ausdruck, in der Programmiersprache LISP eine Folge aus Atomen oder weiteren S-Ausdrücken}*
saddle point Sattelpunkt *m*
saddle-point game Spiel *n* mit Sattelpunkt
saddle-point theorem Sattelpunkttheorem *n*
satisfiability Erfüllbarkeit *f*
satisfice/to nach einer hinreichenden Lösung suchen
satisfy/to befriedigen *{eine zufriedenstellende, aber nicht notwendigerweise optimale Lösung schaffen}*; erfüllen *{z.B. Bedingung}*
saturating integrator Entscheidungsintegrator *m*, inkrementaler Integrierer *m*, Servointegrierer *m*
saturation Sättigung *f* *{ein Bereich, in dem die Ausgangsgröße unabhängig von der weiter zunehmenden Eingangsgröße konstant bleibt}*, Sättigungszustand *m*; Begrenzung *f* *{Übertragungsglied}*

saturation limit Sättigungsgrenze *f*
scalability Skalierbarkeit *f*, Veränderbarkeit *n* des Maßstabes
scale model maßstabgetreues Modell *n*
scaling principle Prinzip *n* der maßstabgerechten Verkleinerung
schedule/to planen
schedule Ablaufplan *m*, Plan *m*, Zeitplan *m*
schema Übersicht *f*, Plan *m*, Schema *n*
scientific and technical production model prognostication wissenschaftlich-technische Prognostizierung *f* von Produktionsmodellen
scientific information wissenschaftliche Information *f*
script 1. Schreibschrift *f*; 2. Skript *n* *{z.B. rahmenartige Struktur für Ereignisfolgen}*
search/to suchen
search Suche *f*, Suchvorgang *m* *{z.B. zur Optimierung}*, Suchlauf *m* *{z.B. bei der Werkzeugwechselsteuerung}*
search cycle Suchzyklus *m*, Suchkreis *m* *{Optimierung}*
search loop Suchschleife *f* *{Rückwärtsoptimierung}*
search operator Suchoperator *m*
search period Suchzeit *f*
search procedure Suchverfahren *n*
search process Suchprozeß *m*
search space Suchraum *m*, Problemraum *m*
search technique Suchverfahren *n*
search tree Suchbaum *m*
searching algorithm Suchalgorithmus *m*
searching cycle Suchzyklus *m*

searching filter Suchfilter *n*
searching frequency Suchfrequenz *f*
searching procedure Suchverfahren *n*
searching process Suchprozess *m*
searching range Suchbereich *m*
searching step Suchschritt *m*
searching strategy Suchstrategie *f*
second-level set Menge *f* zweiter Ordnung (Stufe), Unmenge *f*
second-order differential equation Differentialgleichung *f* zweiter Ordnung
second-order learning Lernen *n* zweiter Ordnung *{Neurocomputer}*
second-order set *s.* second-level set
second-order supercharacter Superzeichen *n* zweiter Ordnung, Super-Superzeichen *n*
secondary sekundär, untergeordnet, Neben-
seed Keim *m*
SELDOM SELTEN *{unscharfer Quantor}*
select/to auswählen *{Wissen}*
selection logic Auswahllogik *f*
selective set Auswahlmenge *f*
self-adaptation Adaptation *f*, Selbstanpassung *f*
self-adaptation by search Selbstanpassung *f* durch Suche, Rückwärtsoptimierung *f*
self-adaptive selbsteinstellend, anpassungsfähig, selbstanpassend, selbstoptimierend, lernend; *{s.a.* adaptive}
self-adjusting selbsteinstellend *{das Netzwerk ist fähig, sich durch Modifizieren seiner internen Prozesse an sich ändernde Inputs anzupassen}*

self-adjustment Selbsteinstellung *f*, Selbstabgleich *m*
self-aligned superintegration logic selbsteingestellte Logik *f* mit höchstem Integrationsgrad, SSL
self-dual Boolean function selbstduale Boolesche Funktion *f*
self-learning computer lernender Computer *m*
self-learning pattern recognition selbstlernende Mustererkennung *f*
self-learning system Lernsystem *n*, lernendes System *n*
self-locking selbstsperrend, selbstsichernd
self-optimization Selbstoptimierung *f*
self-optimizing selbstoptimierend
self-optimizing system selbstoptimierendes System *n*
self-organization Selbstorganisation *f* *{Eigentraining neuronaler Netzwerke}*
self-organizing process selbstorganisierender Prozeß *m*
self-organizing system selbstorganisierendes System *n*
self-programming system selbstprogrammierendes System *n*
self-recording selbstaufzeichnend, selbstregistrierend
self-regulating automatisch, selbsteinstellend
self-regulation Selbstregulation *f*, Selbstregulierung *f*, Selbstausgleich *m*, Ausgleich *m* *{z.B. bei P-Gliedern}*
self-repair Selbstreparatur *f*; *s.a.* self-heal}
self-repairing system selbstreparierendes (selbstheilendes) System *n*

self-reproducing automaton selbstreproduzierender Automat *m*
self-reproduction Selbstreproduktion *f*
self-structurizing selbststrukturierend
self test Selbsttest *m*
semantic semantisch, der Bedeutung nach
semantic analysis semantische Analyse *f*
semantic equivalence semantische Äquivalenz *f* *{Aussagenlogik}*
semantic equivalence criterion semantisches Äquivalenzkriterium *n*
semantic error Semantikfehler *m*
semantic grammar semantische Grammatik *f*
semantic information semantische Information *f*
semantic network semantisches Netz *n* *{Wissensdarstellung mit Hilfe gerichteter Graphen}*
semantic primitive semantisches Primitiv *n*
semantic rule semantische Regel *f*
semantic structure semantische Struktur *f*
semantically complete semantisch vollständig
semantically equivalent semantisch äquivalent *{Aussagenlogik}*
semantics Semantik *f*, Bedeutungslehre *f*
semi-group Halbgruppe *f*
semi-Markov chain Semi-Markow-Kette *f*
semi-Markov process Semi-Markow-Prozeß *m*
semiautomatic halbautomatisch
semicalculable set akzeptierbare Menge *f*
semidynamic system semidynamisches System *n*
semiempirical formula halbempirische Formel *f*
semiempirical model halbempirisches Modell *n*
sending neuron Sendeneuron *n*
sensing Abfühlen *n*, Abtasten *n*, Lesen *n*, Reizaufnahme *f*, Seitenanzeige *f*, Seitenbestimmung *f*
sensitivity curve Empfindlichkeitskurve *f*
sensitivity threshold Empfindlichkeitsschwelle *f*
sensory neuron Sensorneuron *n* *{liefert dem Gehirn Signale von anderen Teilen des Körpers}*
sensory organ Sinnesorgan *n*
sensory robot Sensorroboter *m*
sentence Satz *m*; *{s.a.* logic sentence, declarative sentence*}*
sentence of identity Identitätssatz *m*
sentential calculus Aussagenkalkül *m*
sentential form Aussagenform *f*
separation rule Abtrennungsregel *f* *{Prädikatenlogik}*
seq WENN SO *{Wahrheitsfunktion}*
sequence/to in einer Folge (Reihenfolge) ordnen, anordnen, reihen, ordnen nach natürlichen Zahlen, in die richtige Reihenfolge bringen, der Reihenfolge nach sortieren
sequence Ablauf *m*, Reihenfolge *f*, Schaltfolge *f*, Folge *f* *{Mathematik}*; Sequenz *f* *{Logik}*
sequence-controlled folgegesteuert, programmgesteuert
sequence graph Ablaufgraph *m*

sequence of moves Bewegungsfolge *f*, Bewegungsablauf *m*
sequence of premises Prämissenfolge *f* {*Logik*}
sequence of symbols Zeichenreihe *f* {*z.B. Logik*}
sequence optimization Reihenfolgeoptimierung *f*, Optimierung *f* der Reihenfolge
sequence problem Reihenfolgeproblem *n*
sequence table Schalttabelle *f*, Schaltfolgetabelle *f*
sequenced automation geordnete Automatisierung *f*
sequent decision function sequente Entscheidungsfunktion *f*
sequential sequentiell, aufeinanderfolgend; {*s.a.* serial}
sequential analysis sequentielle Analyse *f*, Sequenzanalyse *f*
sequential circuit sequentielle Schaltung *f*
sequential game Sequentialspiel *n*, sequentielles Spiel *n*
sequential logic sequentielle Logik *f*
sequential logic element logisches Folgeelement *n*
sequential machine sequentieller Automat *m*
sequential network sequentielle Schaltung *f*, Schaltwerk *n*
sequential processing sequentielle (serielle) Verarbeitung *f*, von-Neumann-Verarbeitung *f* {*Gegensatz: Parallelverarbeitung*}
sequential state machine endlicher [sequentieller] Automat *m*, sequentielle Maschine *f*, sequentielles System *n*, Folgeschaltsystem *n*

serial serienmäßig, reihenweise, seriell, zeitparallel; {*s.a.* sequential}
serial computer serieller Computer *m*
serial mode Serienarbeitsweise *f*, serielle Arbeitsweise *f*, Serienbetrieb *m*
serial operation serielle Operation *f*, Serienbetrieb *m*
serial-parallel seriell-parallel
serial-parallel-serial structure Serien-Parallel-Serien-Struktur *f*
serial processing serielle (sequentielle) Verarbeitung *f*
serial queueing system serielles Warteschlangensystem *n*
serial structure Serienstruktur *f*
serial-to-parallel conversion Serien-Parallel-Umsetzung *f*
serializer Parallel-Serien-Umsetzer *m*
series Serie *f*
series operation Serienbetrieb *m*
series-parallel system Serien-Parallel-System *n*
series structure Reihenstruktur *f*
series system Seriensystem *n*
served demand (requirement) bediente Forderung *f* {*Bedienungstheorie*}
server Bedienungsapparat *m*, Bediener *m* {*Bedienungstheorie*}
service in random order Bedienung *f* mit zufälliger Auswahl {*Bedienungssystem*}
service process Bedienungsprozeß *m*
service request Bedien[ungs]anforderung *f*
serving Abfertigung *f* {*Bedienungssystem*}

serving intensity Bedienungsintensität *f* *{Bedienungsprozeß}*
serving process Bedienprozeß *m*, Bedienungsprozeß *m*
servo-controlled robot Roboter *m* mit Servosteuerung
session Dialog *m* *{Mensch-Computer}*, Gespräch *n*
set Menge *f* *{Mengenlehre}*
set algebra Mengenalgebra *f*
set-construction principle Mengenbildungsprinzip *n*
set family Mengenfamilie *f*, Mengensystem *n*
set function Mengenfunktion *f*
set of characteristics Satz *m* von Eigenschaften
set of clauses Klauselmenge *f*
set of conclusions Folgerungsmenge *f*
set of patterns Objektmenge *f* *{Objekterkennung}*
set operation Mengenoperation *f*
set path Sollbahn *f* *{numerische Steuerung}*
set-point value Sollwert *m*, vorgegebener Wert *m*
set power Mengenpotenz *f*
set product Mengenprodukt *n*, Produktmenge *f*, Kreuzmenge *f*
set symbol Mengensymbol *n*
set system Mengensystem *n*, Mengenfamilie *f*
set-theoretical topology mengentheoretische (allgemeine) Topologie *f*
set theory Mengenlehre *f*
set trajectory Solltrajektorie *f*
setting-up mode Einrichtbetrieb *m* *{numerisch gesteuerte Werkzeugmaschine}*
Shannon Shannon *n* *{in der Informationstheorie verwendete Einheit ähnlich dem Bit}*
sharing Teilen *n*, Aufteilen *n*
sharp scharf *{z.B. Modell}*
sheffer function Sheffer-Funktion *f*
shift Verschiebung *f* *{z.B. der Elemente einer Kette}*
shift in zero Nullpunktverschiebung *f*
shift left Linksverschiebung *f*, Verschiebung *f* nach links *{Schieberegister}*
shifting method *s.* penalty function shifting method
short-term memory Kurzzeitgedächtnis *n*, dynamische Wissensbank *f*
shortage Mangel *m*
shortage cost Fehlmengenkosten *pl*, Mangelkosten *pl* *{Lagerhaltungstheorie}*
shortest-path (-route) method Methode *f* des kürzesten Weges, Rundfahrtproblem *n*, Rundreiseproblem *n*
shrink/to schrumpfen, zusammenschrumpfen *{z.B. Kurve zu einem Punkt}*
sigmoid function S-Funktion *f* *{Übertragungsfunktion mit Proportionalbereich und oberer und unterer Sättigungsgrenze zwecks stetigen Übergangs}*
sign function Signumfunktion *f*
signal Signal *n*
signal analysis Signalanalyse *f*
signal element Signalelement *n*
signal pattern Signalmuster *n*
signal set Signalmenge *f*
signal space Signalraum *m*
signal structure Signalstruktur *f*
signal supply Signalvorrat *m*

signature verification Unterschriftsbestätigung *f*
signification Sinn *m*, Bezeichnung *f*, Bedeutung *f*
similar ähnlich
similarity Ähnlichkeit *f*
similarity criterion Ähnlichkeitskriterium *n*
similarity function Ähnlichkeitsfunktion *f*
similarity matrix Ähnlichkeitsmatrix *f*
similarity measure Ähnlichkeitsmaß *n*
similarity principle Ähnlichkeitsprinzip *n*
similarity recognition Ähnlichkeitserkennung *f*
similarity theory Ähnlichkeitstheorie *f*
simple einfach
simple event elementares Ereignis *n*, Elementarereignis *n*
simple function einfache Funktion *f* *{Unterprogramm}*
simple game einfaches Spiel *n* *{Spieltheorie}*
simple group einfache Gruppe *f* *{Gruppentheorie}*
simple inheritance einfache Vererbung *f* *{bei der Übertragung von Eigenschaften hat jede Klasse höchstens eine Oberklasse und beliebig viele Unterklassen}*
simple-jointed chain Kette *f* mit einfachen Gelenken
simplify/to vereinfachen *{z.B. Schaltnetzwerk}*; kürzen, minimieren *{Schaltfunktion}*
simplifying method Kürzungsmethode *f*, Minimierungsmethode *f* *{Schaltfunktion}*
simply connected einfach zusammenhängend *{Punktmenge}*
simulate/to simulieren; nachbilden, modellieren
simulated annealing simuliertes Vergüten *n* *{Glätten der Energiefläche und Schaffung der Möglichkeit zum Verlassen lokaler Minima}*
simulated ironing simuliertes Glätten *n* *{Entfernen von "Falten" in der Energiefläche}*
simulated neuron simuliertes Neuron *n*, Siron *n*
simulation Simulierung *f*, Simulation *f*, Nachbildung *f*, Modellierung *f*
simulation language Simulationssprache *f*
simulation method Simulationsmethode *f*
simulation model Simulationsmodell *n*
simulation parameter Simulationsparameter *m*
simulation system Simulationssystem *n*
simultaneity Gleichzeitigkeit *f*, Simultanität *f*
simultaneous gleichzeitig, simultan
simultaneous operation gleichzeitiges Arbeiten *n*, gleichzeitige Betätigung *f*; Simultanoperation *f* *{Rechenoperation}*; Simultanbetrieb *m*
sine curve man Sinuskurvenmensch *m* *{historische Computergraphik, durch Sinuskurven synthetisiertes Porträt eines Mannes, von Charles Csuri und James Shaffer}*

single-degree-of-freedom system System n mit einem Freiheitsgrad
single-edge graph Einkantengraph m
single-leaf phase plane einblättrige Phasenebene f
single-leaf state plane einblättrige Zustandsebene f
single-leaf surface einblättrige Fläche f
single-level structure Einebenenstruktur f
single-line queue system Warteschlangensystem n mit [nur] einer Leitung
single-line waiting system Wartesystem n mit [nur] einer Leitung *{Bedienungstheorie}*
single-objective function Einzielfunktion f
single-objective structure Einzielstruktur f
single-objective system Einzielsystem n
single-point boundary value problem Ein-Punkt-Randwertaufgabe f
single-server queueing process Bedienungsprozeß m mit einem Bedienungsapparat
single-stage einstufig
single-stage decision einstufige Entscheidung f
single-stage decision process einstufiger Entscheidungsprozeß m
single-state automation Automat m mit [nur] einem Zustand
single-step algorithm Einschrittalgorithmus m
singular singulär
sink Senke f *{z.B. für Daten, s.a.* drain*}*

siron *s.* simulated neuron
site problem Standortproblem n
situation Situation f
situation recognition Situationserkennung f
situation recognition algorithm Situationserkennungsalgorithmus m
sizing robot Sortierroboter m, Roboter m zum Sortieren *{nach Größe}*
slave system untergeordnetes System n
sliced aufgeteilt
sliced microprocessor system Mikroprozessorsystem n mit mehreren aneinandergereihten Mikroprozessoren *{zur Erhöhung der Verarbeitungsbreite}*
slot Slot n, Fach n *{Rahmenelement, das mit Informationen zu einem speziellen Sachverhalt gefüllt wird}*
small knowledge system kleines Wissenssystem n *{System mit weniger als etwa 500 Regeln}*
small-scale similarity Ähnlichkeit f im Kleinen
small-variations method Methode f der kleinen Variationen
smallest-parameter perturbation method Methode f des kleinsten Parameters *{Methode von Poincare}*
smart card intelligente Leiterplatte f, intelligente Karte f
smart terminal intelligentes Terminal n *{Terminal mit zusätzlichen Eigenschaften, z.B. visuelle Hervorhebungen, Bildschirmspeicher u.a., jedoch keine Programmierbarkeit mittels RAM}*
soft restriction weiche Beschränkung f, weiche Restriktion f *{Optimierung}*

SOL *s.* second-order learning
solidus Soliduslinie *f*
solute/to lösen *{z.B. Gleichung}*
solution algorithm Lösungsalgorithmus *m*
solution approach Lösungsweg *m*
solution graph Lösungsgraph *m*
solution model Lösungsmodell *n*
solution path Lösungspfad *m*, Lösungsweg *m {erfolgreicher Pfad durch den Suchraum}*
solution tree Lösungsbaum *m*
solvable lösbar *{Gleichungssystem}*
solve/to lösen *{z.B. eine Aufgabe}*
solve a problem/to ein Problem lösen
solved conflict gelöster Konflikt *m*
solver *s.* problem solver
solving Lösen *n*
soma Soma *n {Stelle an einem Neuron, wo eintreffende Signale addiert werden}*
sophisticated ausgeklügelt, kompliziert; hochentwickelt
sort algorithm Sortieralgorithmus *m*
sorting tree Sortiernetzwerk *n*, Sortierbaum *m*
soundness Korrektheit *f {z.B. in der Logik}*
source Quelle *f {z.B. Wissensbasis}*
source language Ursprungssprache *f*, Quellsprache *f*
source-language debugging Fehlerbeseitigung *f* in Primärsprache
source-language translation Primärsprachenübersetzung *f*
source library Quellbibliothek *f*
source node Quellknoten *m*
source set Quellsatz *m*
source traffic Quellenverkehr *m*, Engset-Verkehr *m*

space of characteristics Merkmalsraum *m*
space of objects Objektraum *m*
space structure Raumstruktur *f*
spatial räumlich
spatial pattern räumliches Muster *n {parallele Signalwerte zu einer gegebenen Zeit}*
speaker-dependent recognition sprecherabhängige Spracherkennung (Erkennung) *f {nach Lernprozeß}*
speaker-dependent recognition system sprecherabhängiges Erkennungssystem *n*
speaker-dependent speech recognition sprecherabhängige Spracherkennung *f*
speaker-independent recognition sprecherunabhängige Spracherkennung (Erkennung) *f*
speaker-independent speech recognition sprecherunabhängige Spracherkennung *f*
special function spezielle Funktion *f*
special-purpose logic Speziallogik *f*
special[-type] robot Spezialroboter *m*, Einzweckroboter *m*
spectrum logic Spektrallogik *f*
speech Sprache *f*
speech analysis Sprachanalyse *f*
speech analyzer Sprachanalysator *m*
speech communication Sprachkommunikation *f*
speech compression Sprachkompression *f*
speech input Spracheingabe *f*
speech module Sprachmodul *m*, Sprachbaustein *m*
speech output Sprachausgabe *f*

speech recognition Sprach[wieder]erkennung f, Sprachverstehen n *{maschinelle Wahrnehmung von Sprache}*
speech recognition system Spracherkennungssystem n
speech scrambler Sprachverstümmler m
speech simulation Sprachsimulation f
speech synthesis Sprachsynthese f
speech understanding Sprachverständnis n
spelling checker Rechtschreibprüfprogramm n, Prüfprogramm n für Rechtschreibung *{Textverarbeitung}*
spelling-error detection Auffindung f von Rechtschreibfehlern
spherical coordinate robot Kugelkoordinatenroboter m *{Roboter mit Effektorbewegung auf vorgegebener Kugelfläche}*
spherically jointed robot Kugelgelenkroboter m, Roboter m mit Kugelgelenken
split screen Mehrfelderbildschirm m, aufgeteilter Bildschirm m
spot welding robot Punktschweißroboter m, Roboter m zum Punktschweißen
spray head Sprühkopf m *{z.B. beim Farbsprühroboter}*
spray-painting robot Farbspritzroboter m, Industrieroboter m zum Farbspritzen
stability in the large Stabilität f im Großen
stability in the small Stabilität f im Kleinen
stable stabil, beständig, konstant
stable nodal point stabiler Knotenpunkt m
stable state stabiler Zustand m
stable system stabiles System n
standard sum disjunktive Normalform f *{Schaltalgebra}*
starting event Startereignis n, Anfangsereignis n *{Graph}*
starting function Anfangsfunktion f, Ausgangsfunktion f
starting menu Startmenü n
starting solution Ausgangslösung f, Anfangslösung f *{lineare Optimierung}*
state curve Zustandskurve f
state-dependent zustandsabhängig
state equation Zustandsgleichung f
state estimation Zustandsschätzung f
state graph Zustandsgraph m *{Graph, bei dem die Knoten den Systemzustand und die Verbindungslinien Operatoren von einem Zustand in einen anderen repräsentieren}*
state-independent zustandsunabhängig
state-independent partition zustandsunabhängige Partition f
state-linear zustandslinear
state number minimization of an automaton Minimierung f der Anzahl der Zustände eines Automaten
state observer Zustandsbeobachter m *{Systemtheorie}*
state of organization Organisiertheit f
state of the art Stand m der Technik
state plane Zustandsebene f
state probability Zustandswahrscheinlichkeit f

state recognition Zustandserkennung *f*
state reduction Zustandsreduktion *f*
state set Zustandsmenge *f*
state space Zustandsraum *m*
state space analysis Zustandsraumanalyse *f*
state space method Zustandsraummethode *f*
state space trajectory Zustandsraumtrajektorie *f*, Trajektorie *f* im Zustandsraum
state table Zustandstabelle *f*, Automatentabelle *f*
state trajectory Zustandsbahn *f*, Zustandstrajektorie *f*
state transition matrix Zustandsübergangsmatrix *f*
state value Wert *m* der Zustandsgröße (Zustandsvariablen)
state variable Zustandsgröße *f*, Zustandsvariable *f*, Zustandsveränderliche *f*
state variable equation Zustandsgleichung *f*, Zustandsgrößengleichung *f*
state variable vector Zustandsgrößenvektor *m*
statement 1. Erklärung *f*, Anweisung *f*; 2. Aussage *f* *{Logik}*; 3. Forderung *f*
static statisch
  static state statischer Zustand *m*, Ruhezustand *m*, Gleichgewichtszustand *m*
stationary stationär, ortsfest; eingeschwungen, ruhend *{s.a. steady-state}*
statistical game statistisches Spiel *n*, Spiel *n* gegen die Natur
statistical hypothesis statistische Hypothese *f*
statistical model statistisches Modell *n*
statistics Statistik *f*
status vector Zustandsvektor *m*
steady 1. stetig; gleichmäßig; 2. gleichförmig, konstant
steady-state condition Gleichgewichtsbedingung *f*, Bedingung *f* für das zeitliche Gleichgewicht, Stationaritätsbedingung *f*
steady-state optimization statische Optimierung *f*, Gleichgewichtsoptimierung *f*
steady-state optimizing problem statisches Optimierungsproblem *n*
steady-state output stationäre Ausgangsgröße *f*, stationärer Lösungsteil *m* der Ausgangsgröße
steady-state process stationärer Prozeß *m*
steady-state simulation statische Simulierung *f*
steady-state solution statische (stationäre) Lösung *f*, Lösung *f* im statischen (eingeschwungenen) Zustand
step Stufe *f*, Schritt *m*, Sprung *m* *{Signalverlauf}*
step-extremal system Schrittextremalsystem *n*
step frame Schrittraster *m*
step gain Stufengewinn *m* *{dynamische Optimierung}*
step process Stufenprozeß *m* *{dynamische Optimierung}*
step vector Schrittvektor *m* *{Suchschritt}*
step width Schrittweite *f* *{Suchverfahren}*

stepping stone method Stepping-stone-Methode *f*, Primal-Dual-Algorithmus *m* *{lineare Optimierung}*
stereotyped situation stereotype Situation *f* *{sich wiederholende, generische Situation}*
stimulus Stimulus *m*, Reiz *m* *{Zustand des Inputs eines Systems}*
stimulus equivalence Reizäquivalenz *f*
stimulus response model Reizreaktionsmodell *n* *{Black-box-Methode}*
stochastic stochastisch, regellos, zufällig, nichteindeutig, probabilistisch *{s.a.* random*}*
stochastic automaton stochastischer Automat *m*
stochastic behaviour stochastisches Verhalten *n*
stochastic game stochastisches Spiel *n*
stochastic optimization stochastische Optimierung *f*
stochastic quasi-gradient method stochastische Quasigradientenmethode *f*
stochastic simulation stochastische Simulation *f*
stochastic system stochastisches (zufälliges) System *n*
stochastic weight decay stochastische Gewichtsverringerung *f*
stock control Lagerüberwachung *f*, Lagerhaltungsüberwachung *f*
stock-keeping cost Lagerkosten *pl*, Lagerhaltungskosten *pl* *{Lagerhaltungstheorie}*
stock-keeping model Lagerhaltungsmodell *n*
stock-keeping problem Lagerhaltungsproblem *n*
stock-keeping system Lagerhaltungssystem *n*
stock-keeping theory Lagerhaltungstheorie *f*
stocking Lagerhaltung *f*
stopword unwesentliches Wort *n* [in einem Kontext] *{Wort mit keinem oder nur geringem Informationsgehalt, z.B. der, die, das, von, aus etc.}*
storage-and-information-retrieval system System *n* zum Speichern und Wiederauffinden von Informationen
stored logic gespeicherte Logik *f*
stored-logic computer Automat *m* mit gespeicherter Logik
straight endpoint-to-point control Punkt-Strecken-Steuerung *f*
strategic behaviour strategisches Verhalten *n*
strategic case strategische Situation *f*
strategic computer Strategiecomputer *m*
strategic decision strategische Entscheidung *f*
strategic equivalence strategische Äquivalenz *f*, Äquivalenz *f* von Spielen
strategic game strategisches Spiel *n*
strategic knowledge strategisches Wissen *n*
strategic model strategisches Modell *n*
strategic situation strategische Situation *f*
strategy Strategie *f*
strategy computer Strategiecomputer *m*

strategy of game Spielstrategie *f*, Strategie *f* des Spiels
strategy planning Strategieplanung *f*
strategy set Strategiemenge *f*
strategy space Strategieraum *m*
stratified language geschichtete Sprache *f*
stream Strömung *f*, Strom *m*
stream intensity Stromintensität *f* {*Ankunftsprozeß*}
stream of demands (requirements) Forderungsstrom *m* {*Bedienungstheorie*}
strength Stärke *f* {*einer Regel*}
strengthen/to verstärken {*Reiz*}
stretch/to aufspannen {*Graph*}
stretched graph aufgespannter Graph *m*
strict implication strikte Implikation *f* {*Logik*}
stringent scharf {*z.B. Stabilitätsbedingungen*}
strong connection starker Zusammenhang *m* {*Graph*}
strong convergence starke Konvergenz *f*
strong extreme value starkes Extremum *n*
strongly connected graph stark zusammenhängender Graph *m*
structurable strukturierbar
structural strukturell, Struktur-
structural ambiguity strukturelle Mehrdeutigkeit *f*
structural analogy Strukturanalogie *f*, strukturelle Analogie *f*
structural analysis Strukturanalyse *f*
structural attribute Strukturattribut *n*
structural automata theory Strukturtheorie *f* von Automaten
structural equivalence Strukturäquivalenz *f*, strukturelle Äquivalenz *f*
structural instability Strukturinstabilität *f*
structural level Strukturebene *f*
structural matrix Strukturmatrix *f*
structural model strukturelles Modell *n*
structural optimization Strukturoptimierung *f*, strukturelle Optimierung *f*, Tragwerksoptimierung *f* {*Bauwesen*}
structural redundancy Strukturredundanz *f*, strukturelle Redundanz *f*
structural semantics strukturelle Semantik *f*
structural stability Strukturstabilität *f*, strukturelle Stabilität *f*
structural synthesis Struktursynthese *f*
structural theory of automata strukturelle Automatentheorie *f*
structural variation Strukturveränderung *f*, strukturelle Veränderung *f*
structurally instable system strukturinstabiles System *n*
structurally stable system strukturstabiles System *n*
structurate/to strukturieren
structure Struktur *f*, Aufbau *m*, Organisation *f* {*z.B. eines Systems*}
structure index Strukturindex *m*
structure instability Strukturinstabilität *f*
structure matrix Strukturmatrix *f*
structure model Strukturmodell *n*

structure optimization Strukturoptimierung *f*
structure parameter Strukturparameter *m*
structure problem Strukturproblem *n*
structure rule Strukturregel *f*
structure simulator Struktursimulator *m*
structure theory Strukturtheorie *f*
structured strukturiert
structured set strukturierte Menge *f*
subclass Unterklasse *f*, Teilklasse *f* *{Mengenlehre}*
subcontrol Abwärtssteuerung *f* *{lernender Automat}*
subdivided system unterteiltes System *n*
subdivision Unterteilung *f*
subfield Unterkörper *m*, Teilkörper *m* *{Mathematik}*
subgoal Unterziel *n*, untergeordnetes Ziel *n*, Teilziel *n*, Zwischenziel *n* *{bei einer Problemlösung notwenigerweise zu erreichendes Ziel, ehe das eigentliche Ziel erreicht werden kann}*
subgraph Subgraph *m*, Teilgraph *m*
subgroup Untergruppe *f*
subject Subjekt *n*
subject-oriented programming system sachgebietsorientiertes Programmiersystem *n*, SOPS
submatrix Untermatrix *f*, Submatrix *f*
submodel Submodell *n*, Untermodell *n*
suboptimal *s.* suboptimum
suboptimization (suboptimizing) Suboptimierung *f*
suboptimum suboptimal
suboptimum solution suboptimale Lösung *f*
suboptimum strategy suboptimale Strategie *f*
subordinated system untergeordnetes System *n*
subplan Teilplan *m* *{Plan für eine Teillösung des Problems}*
subproblem Teilproblem *n* *{muß z.B. vor der Lösung des Gesamtproblems gelöst werden}*
subset 1. Geräteteil *m*; Baugruppe *f*; 2. Untermenge *f*, Teilmenge *f* *{Mengenlehre}*
subspace method Hilfsraummethode *f* *{ein Musterkennungsalgorithmus}*
substitute by/to ersetzen durch, austauschen durch (mit)
substitution Substitution *f*, Austauschen *n*, Ersatz *m*
substitution axiom Ersetzungsaxiom *n*
substitution error Substitutionsfehler *m*
subsystem untergeordnetes System *n*, Subsystem *n*, Untersystem *n*, Teilsystem *n* *{eines größeren Systems}*
subsystem optimization Optimierung *f* von Teilsystemen
subsystem structure Teilsystemstruktur *f*
subtask Unteraufgabe *f*
subvector space Untervektorraum *m*
succeeding activity nachfolgende Aktivität *f* *{Netzplantechnik}*
success Erfolg *m*
success balancing Erfolgsbilanzierung *f*
success criterion Erfolgskriterium *n*
success function Erfolgsfunktion *f*
sufficient condition hinreichende Bedingung *f*

sufficient reason rule Satz *m* vom zureichenden Grund
summation function Summationsfunktion *f* *{eine Aktivierungsfunktion, bei der Signale addiert und das Ergebnis mit einem inneren Schwellenwert verglichen wird}*
supercharacter Superzeichen *n*
superclass Oberklasse *f*
supercontrol Aufwärtssteuerung *f* *{lernender Automat}*
superfluous information überflüssige Information *f*
supergroup Obergruppe *f*, Übergruppe *f*
supervised learning überwachtes Lernen *n* *{äußere Beeinflussung des Netzwerkes in Abhängigkeit von der Richtigkeit des Outputs und der Gewichte}*
sure event sicheres Ereignis *n*
surface knowledge Oberflächenwissen *n*, Erfahrungswissen *n*, heuristisches Wissen *n*
surjective function Surjektion *f* *{Mengenlehre}*
surprise content Überraschungsgehalt *m* *{Information}*
survivability Überlebensfähigkeit *f* *{z.B. von Einrichtungen}*
surviving probability Überlebenswahrscheinlichkeit *f* *{Zuverlässigkeitstheorie}*
surviving probability distribution Überlebenswahrscheinlichkeitsverteilung *f*
switch matrix Schaltmatrix *f*
switching Schaltvorgang *m*; Schalt-
switching algebra Schaltalgebra *f*
switching circuit logische Schaltung (Verknüpfungsschaltung) *f*,
Schaltkreis *m*, Schaltsystem *n*, Schaltnetzwerk *n*, Schaltnetz *n*
switching element Schaltglied *n*, Schaltelement *n*, Verknüpfungsglied *n*, [logisches] Verknüpfungselement *n*
switching function Schaltfunktion *f*, binäre logische Funktion *f*
switching system Schaltsystem *n*
switching theory Theorie *f* der Schaltnetzwerke, Schalttheorie *f*
switching variable Schaltvariable *f*
syllogism Syllogismus *m* *{deduktive Aussagenverknüpfung mit zwei Prämissen}*
syllogistics Syllogistik *f*
symbol Symbol *n*, Zeichen *n*, Schaltzeichen *n*, Sinnbild *n*
symbol dictionary Symboltabelle *f*
symbol library Symbolbibliothek *f*
symbol processing Symbolverarbeitung *f*
symbol sequence Symbolfolge *f*
symbol string Symbolkette *f*
symbol supply Symbolvorrat *m*
symbol table Symboltabelle *f*
symbol variable Symbolvariable *f*
symbolic concordance symbolische Übereinstimmung *f*
symbolic language Symbolsprache *f*
symbolic listing Symbolliste *f*
symbolic logic formale (symbolische) Logik *f*, Symbollogik *f* *{s.a. mathematical logic}*
symbolic model formales (symbolisches) Modell *n*
symbolic name symbolischer Name *f*, symbolische Bezeichnung *f*
symbolic notation symbolische Schreibweise (Notation) *f*

symbolic parameter symbolischer Parameter *m*
symbolic process symbolischer Prozeß *m*
symbolic programming Symbolverarbeitung *f*
symbolic representation sinnbildliche (symbolische) Darstellung *f*
symbolic sentence logische Aussage *f*
symbolic unit symbolische Einheit *f*
symbolics Symbolik *f {Lehre von der Substitution konkreter Objekte durch abstrakte Darstellungen}*
symbolics computation Symbolverarbeitung *f*
symbolize/to symbolisieren, versinnbildlichen
symbology Symbolik *f*
symmetric Boolean function symmetrische Boolesche Funktion *f*
symmetric difference symmetrische Differenz *f {Mengenlehre}*
symmetric group symmetrische Gruppe *f*
symmetric matrix symmetrische Matrix *f*
symmetric relation symmetrische Relation *f*
symmetrization Symmetrierung *f*
symmetrize/to symmetrieren
synapse Synapse *f {Bereich des elektrochemischen Kontaktes zwischen Neuronen}*
synchronous automaton synchroner Automat *m*
synchronous logic synchrone Logik *f*
synergetics Synergetik *f*

synoptic synoptisch, umfassend
syntactic analyzer Syntaxanalysator *m*
syntactic controlled translator syntaxgesteuerter Übersetzer *m*
syntactic error Syntaxfehler *m*
syntactic program checking Syntaxkontrolle *f*
syntactical analysis syntaktische Analyse *f*
syntactically complete syntaktisch vollständig
syntagmatic relation syntagmatische Relation *f*
syntax Syntax *f*, Satzlehre *f {Aufbau einer Sprache, z.B. einer Programmsprache}*
syntax checker Syntaxprüfer *m*
synthesis Synthese *f {z.B. Sprachsynthese}*
synthesis of automata Synthese *f* von Automaten
synthesis of discrete automata Synthese *f* diskreter Automaten
synthesis of speech signals Sprachsignalsynthese *f*
synthesis problem Syntheseproblem *n*
synthesis test Synthesetest *m*
synthetic language synthetische Sprache *f*
synthetic relation synthetische Relation *f*
synthetizable synthetisierbar
system System *n*
system analysis Systemanalyse *f*
system approach Systemlösung *f*
system class Systemklasse *f*
system component Systembestandteil *m*
system concept Systemkonzept *n*

system consultation Systemabfrage *f* {*Expertensystem*}
system dependency Systemabhängigkeit *f*, Abhängigkeit *f* vom System
system-dependent systemabhängig, anlagenabhängig
system description Systembeschreibung *f*
system design Systementwurf *m*
system-dynamic systemdynamisch
system effectiveness Wirksamkeit *f* eines Systems, Systemwirksamkeit *f*
system environment Systemumgebung *f*
system-environment communication System-Umwelt-Kommunikation *f*, Kommunikation *f* zwischen System und Umwelt
system failure Systemausfall *m*, Ausfall *m* des Systems, Ausfall *m* eines Regelkreises, Ausfall *m* einer Regelstrecke (Strecke) *f*; Anlagenausfall *m*
system hierarchy Systemhierarchie *f*
system identification Systemidentifikation *f*, Systemidentifizierung *f*
system management Systemverwaltung *f*
system matrix Systemmatrix *f*
system model Systemmodell *n*, Modell *n* des Systems
system model[l]ing Systemmodellierung *f*, Modellierung *f* von Systemen
system optimization Systemoptimierung *f*
system performance Systemverhalten *n*
system reliability Systemzuverlässigkeit *f*, Zuverlässigkeit *f* des Systems; Anlagenzuverlässigkeit *f*
system stability Stabilität *f* eines Systems
system state Systemzustand *m*
system structure Systemstruktur *f*
system synthesis Systemsynthese *f*
system theory Systemtheorie *f*
system-to-system communication Dialog *m* zwischen Computern
system with distributed parameters System *n* mit verteilten Parametern
system with memory speicherbehaftetes System *n*
system with self-healing selbstheilendes (selbstreparierendes) System *n*
system with several degrees of freedom System *n* mit mehreren Freiheitsgraden
systematic systematisch
systolic array systolisches Array *n*

T

tactical planning taktische (kurzzeitige) Planung *f*
tactics Taktik *f*
tactile recognition capability Tastfähigkeit *f*, Tastsinn *m* {*Roboter*}
tactile sense Tastsinn *m*
tactile sensing Berührungsabtastung *f*
tag/to markieren
tailored auf den Anwendungsfall zugeschnitten, problemorientiert, maßgeschneidert

tailoring problem Zuschnittproblem *n*
tampering unbefugter Eingriff *m*
target 1. Ziel *n {z.B. bei der Optimierung}; s.a. objective*; 2. Speicherplatte *f*, Signalplatte *f {Speicherröhre}*; 3. Target *n {z.B. für ionisierende Strahlung}*
target function Zielfunktion *f*
target language Zielsprache *f*
target level Zielebene *f*, Zielniveau *n*
target phase Zielphase *f*
target program Objektprogramm *n*
target set Zielsatz *m*
task Aufgabe *f*, Arbeitsaufgabe *f*; Prozeß *m*, Rechenprozeß *m*, Programmabschnitt *m*, Aufgabe *f {Schritt eines Jobs}*
task-oriented language aufgabenorientierte Sprache *f*, Objekt-sprache *f*
task queue Aufgabenwarteschlange *f*
tautological aussagenlogisch wahr, tautologisch
tautologism Tautologismus *m*
tautology Tautologie *f*
taxonomy Taxonomie *f {hierarchische Begriffsdarstellung}*
Taylor series expansion of Boolean functions Taylor-Entwicklung *f* Boolescher Funktionen
teach/to lehren
teach in/to lehren, belehren *{lernender Automat}*; einlesen *{z.B. Bewegungsablauf}*
teach the robot a new job/to den Roboter eine neue Arbeitsaufgabe lehren
teach-in Programmiereingabe *f* durch Lehren *{z.B. beim Industrieroboter}*

teach-in method Teach-in-Verfahren *n {Programmieren von NC-Maschinen}*
teaching automaton (computer) Lehrautomat *m*, Lehrmaschine *f*
teaching in pattern recognition Belehrung *f* in der Mustererkennung
teaching pattern Lehrmuster *n*
teaching program Lehrprogramm *n*
teaching step Lehrschritt *m*
teaching system Lehrsystem *n*
teaching tool Lehrhilfe *f*
teachware Lern-Software *f*, Ausbildungs-Software *f*, Teachware *f {z.B. Lehrprogramm für Softwareanwendungen}*
technical cybernetics technische Kybernetik *f*
teleoperator Teleoperator *m*, Fernbedienungseinrichtung *f {am Roboter}*
temperature Temperatur *f {im Zusammenhang mit dem Verhalten neuronaler Netzwerke eine explizit zeitlich abnehmende Funktion}*
tensor Tensor *m {Vektor}*
term 1. Term *m {kleinster Teil eines Ausdruckes, dem ein Wert zugeordnet werden kann}*; 2. Glied *n {mathematische Funktion, Reihenentwicklung}*
term discrimination Begriffsunterscheidung *f*
terminal analysis [of a system] äußere Systemanalyse (Prozeßanalyse) *f {Analyse unter Betrachtung des Systems als black box mit Hilfe der Eingangs- und Ausgangsgrößen}*
terminal node Blatt *n {eines Baumes}*, letzter (terminaler, nicht

termination

weiter zerlegbarer) Knoten *m* *{Suchbaum}*
termination Ende *n*, Abbruch *m*
terminological check terminologische Kontrolle *f*
ternary dreiwertig, trivalent, dreistellig *{Relation}*; ternär *{Logik}*
ternary logic ternäre Logik *f*, Dreizustandslogik *f*
ternary switching algebra ternäre Schaltalgebra *f*
tertium non datur s. excluded-third rule
test/to testen, durch Eliminination schlußfolgern *{aus mehreren angebotenen Lösungen werden solche entfernt, die bestimmte Kriterien nicht erfüllen}*
THEN DANN
THEN AND ONLY THEN GENAU DANN WENN *{Logik}*
THEN IF DANN WENN *{Logik}*
theorem Theorem *n*, hypothetische Schlußfolgerung *f* *{Aussage, die auf der Grundlage vorhandener Prämissen noch bewiesen werden soll}*
   theorem proving Beweisen *n* von Theoremen *{Problemlösung mittels deduktiver Logik}*
theoretician Theoretiker *m*
theoretic[al] theoretisch
theory Theorie *f*
   theory of automata Automatentheorie *f*
   theory of finite automata Theorie *f* endlicher Automaten
   theory of games Spieltheorie *f*
   theory of logical nets Theorie *f* der logischen Netze
   theory of random processes Theorie *f* der Zufallsprozesse
   theory of similarity Ähnlichkeitstheorie *f*
   theory of the optimum filter Theorie *f* der optimalen Filter
therapy Therapie *f*
THERE IS ES GIBT EIN *{Prädikatenlogik}*
third-generation computer Computer *m* der dritten Generation
thirty-two bit system 32-Bit-System *n*
thought Gedanke *m*
thought experiment Gedankenversuch *m*
thought processing Gedankenverarbeitung *f*, Verarbeitung *f* von Gedanken
three-houses-and-three-wells problem Problem *n* der drei Häuser und drei Brunnen, Problem *n* der neun Fußwege, Problem *n* der zänkischen Nachbarn *{Graphentheorie}*
three-layer perceptron Dreiebenenperceptron *n*
three-level network Dreiebenennetzwerk *n*
three-person game Dreipersonenspiel *n*
three-state logic Dreizustandslogik *f*, Logik *f* mit drei Zuständen, ternäre Logik *f*
three-valued logic dreiwertige Logik *f*
threshold Schwell[en]wert *m* *{innerer Aktivierungswert in einem Neuron, der den Output umschaltet, sobald der Wert erreicht ist}*
threshold function Schwell[en]wertfunktion *f* *{Übertragungsfunktion mit Ja/Nein-Antwort}*
threshold value Schwell[en]wert *m*

time behaviour Zeitverhalten *n*, zeitliches Verhalten *n*, Zeitverlauf *m*
time coordinate Zeitkoordinate *f*, zeitliche Koordinate *f*
time-independent zeitunabhängig
time scaling 1. Zeitraffung *f*; 2. Zeitdehnung *f*
time scheduling Zeit[ablauf]planung *f*
time sharing Zeitteilbetrieb *m* {gleichzeitige Benutzung eines Computers durch mehrere, voneinander unabhängige Programme}
time slice Zeitscheibe *f*
time system Laufzeitsystem *n*
timing Zeitverhalten *n*, Zeitablauf *m*, [vorgegebene] Zeitfolge *f*, Taktgabe *f*, Zeitsteuerung *f*, Taktzeit *f* {s.a. time ...}
tool Hilfsmittel *n*; Werkzeug *n*; Tool *n* {z.B. Spezialsoftware}
top-down approach Rückwärtsverkettungsmethode *f* {erwartungsgeführte Problemlösungsmethode}
top-down logic Top-Down-Logik *f*
top-down method Top-down-Methode *f*, Methode *f* des Hypothetisierens und Testens (erwartungsgesteuerte oder zielgerichtete Methode zur Problemlösung, die auf Modellen und anderem Wissen basiert)
topology 1. Topologie *f* {Mathematik}; 2. Lage *f* {von Elementen in integrierten Schaltungen}
total vollständig, gesamt, total
total black box totale Black-box *f*
total malfunction Gesamtausfall *m*, Totalausfall *m* {s.a. total failure}
total ordering relation Vollordnungsrelation *f*
total system Gesamtsystem *n*
totality Gesamtheit *f*
touch sensor Berührungssensor *m*
touring problem Tourenproblem *n* {lineare Optimierung}
tournament Turnier *n* {Graphentheorie}
tournament sink Senke *f* des Turniers
tournament source Quelle *f* des Turniers
traffic Verkehr *m* {z.B. Bedienungstheorie}
traffic analysis Verkehrsanalyse *f*
traffic control computer Verkehrssteuercomputer *m*, Verkehrsleitcomputer *m*
traffic value Verkehrswert *m* {Bedienungstheorie}
trainable lernfähig
trainable system lernender (lernfähiger) Automat *m*
training software Schulungsprogramm *n*
trajectory Trajektorie *f*, Bewegungsbahn *f*, Bahn *f*, Zustandsbahn *f*, Flugbahn *f*
transaction processing Teilhaberbetrieb *m*, Dialogverarbeitung *f*
transfer function Übertragungsfunktion *f*
transinformation Transinformation *f*, Wechselinformation *f*
transinformation content Transinformationsgehalt *m*
transition graph Übergangsgraph *m*
transitive transitiv
transitive dependence transitive Abhängigkeit *f*
transitivity Transitivität *f*
translation Übersetzung *f*, Umwand-

translator 116

lung *f*, Umsetzung *f*, Konvertierung *f* {*z.B. von Daten*}
translation language Übersetzungssprache *f*
translator Zuordner *m* {*Informationen*}; Übersetzer *m*, Übersetzungsprogramm *n*
transport Transport *m*
transport network Transportnetz *n* {*Graph*}
transport optimization Transportoptimierung *f*
transport problem Transportproblem *n*, Verteilungsproblem *n*, Distributionsproblem *n*
transportation optimization Transportoptimierung *f*
transportation problem Transportproblem *n*, Verteilungsproblem *n*, Distributionsproblem *n*
trapdoor function Falltürfunktion *f*
travelling-salesman problem Handelsreisendenproblem *n*, Problem *n* des Handelsreisenden, Sammelfahrtproblem *n*
tree Baum *m* {*Struktur, z.B. Suchbaum*}
tree-like baumähnlich
tree-like process verzweigter Prozeß *m*
tree representation Baumdarstellung *f*
tree search Baumsuche *f*
tree-search technique Baumsuchverfahren *n*
tree structure Baumstruktur *f* {*spezieller Graph, bei dem ein Knoten, die Wurzel, keinen Vorgänger hat*}
tree-structured directory Adreßbuch *n* mit Baumstruktur

trend Trend *m*
trial Versuch *m*, Erprobung *f*
trial-and-error method Versuch-und-Irrtum-Methode *f*, Trial-and-error-Methode *f*, Methode *f* des Lernens durch Erfolg
trimming problem Zuschnittproblem *n*, Abgleichproblem *n*
trouble Störung *f*
trouble diagnosis Störungsdiagnose *f*
trouble-location problem Störungsortungsproblem *n*
trouble-shooting tree Fehlersuchbaum *m*
trouble tracking Fehlersuche *f*
true wahr {*Logik*}
true conclusion wahre Schlußfolgerung (Konklusion) *f*, wahrer Schluß *m*
true [declarative] sentence wahre Aussage *f*
true signal H-Signal *n*, "WAHR"-Signal *n*
truncation Diskretisierung *f*, Quantisierung *f*; Rundung *f*, Abrundung *f* {*Zahlen*}
truncation error Quantisierungsfehler *m*, Diskretisierungsfehler *m*, Fehler *m* bei der Diskretisierung analoger Werte, Abbruchfehler *m*, Abschneidefehler *m*, Rundungsfehler *m*
trunk Kanal *m*, Übertragungsweg *m*; Vielfachleitung *f*, Hauptleitung *f*
truth concept Wahrheitsbegriff *m*, Wahrheitskonzept *n*
truth function Wahrheits[werte]funktion *f*
truth maintenance Wahrheitserhaltung *f* {*Problemlösungsmethode, bei*

der während der Lösung gewonnene Erkenntnisse und Überzeugungen so verfolgt werden, daß sie, falls fehlerhaft, bei späterem Auftreten von Widersprüchen zurückgenommen werden}
truth table Funktionstafel (Funktionstabelle) *f* für Wahrheitswerte, Boolesche Operationstafel *f*, Wahrheitstafel *f*, Wahrheitswertetafel *f*, Wahrheitstabelle *f*, Wahrheitswertetabelle *f*; Schaltbelegungstabelle *f*
truth table generator Wahrheitswertetafelgenerator *m*
truth threshold Wahrheitsschwelle *f {Expertensystem}*
truth value Wahrheitswert *m {z.B. einer der beiden möglichen Werte in der binären Logik}*
truth-value analysis Wahrheitswerteanalyse *f {Logik}*
truth variable Wahrheitswertevariable *f*
try/to versuchen, erproben, ausprobieren; prüfen
try Versuch *m {z.B. zur Auffindung einer Information}*
trying automaton probierender Automat *m*
tse Tse *n {chinesischer Buchstabe, an dem erstmalig die Bilderkennung mittels Tse-Computer erprobt wurde}*
tse computer Tse-Computer *m {Bilderkennung}*
tse detection Bildabtastung *f* mit Tse-Computer
tse element Tse-Element *n {Tse-Computer}*
tse network Tse-Schaltung *f {Tse-Computer}*

Turing-decidable predicate Turing-entscheidbares Prädikat *n*
Turing-enumerable relation Turing-aufzählbare Relation *f*
Turing machine Turing-Automat *m*, Turing-Maschine *f*
Turing table Turing-Tafel *f*
turn Windung *f*; Umdrehung *f*; Umlauf *m {Graph}*
tutor system Lehrsystem *n*, Tutorsystem *n {künstliche Intelligenz}*
two-input identity circuit Identitätsschaltung *f* (Identitätsgatter *n*, Identitätselement *n*, Identitätsglied *n*) mit zwei Eingängen
two-input OR gate ODER-Gatter *n* mit zwei Eingängen
two-input quad gate Vierfachgatter *n* mit je zwei Eingängen
two-level system Zweiebenensystem *n {hierarchiches System}*
two-loop adapt[at]ion Zweischleifenadaption *f*, zweischleifige Adaption *f*
two-matrices game Bimatrixspiel *n*
two-person game Zweipersonenspiel *n*
two-person non-zero sum game Zweipersonen-Nichtnullsummenspiel *n*
two-person zero-sum game Zweipersonen-Nullsummenspiel *n*, Matrixspiel *n*
two-phase queueing system Zweiphasenbedienungssystem *n*, Tandembedienungssystem *n*
two-sort transportation problem Zweisortentransportproblem *n*
two-state system System *n* mit zwei [stabilen] Zuständen

two-value principle Prinzip *n* der Zweiwertigkeit *{Aussagenlogik}*
two-valued function zweiwertige (Boolesche) Funktion *f*
two-valued logic zweiwertige (binäre) Logik *f*
typical set typische Menge *f*

U

unary operation unäre Operation *f*
unary operator unärer Operator *m*
unary predicate einstelliges Prädikat *n* *{Prädikatenlogik}*
unbounded domain unbegrenzter Bereich *m*
unbounded optimization problem Optimierungsproblem *n* ohne Schranken
uncertainty Ungewißheit *f*
unconditional expectation unbedingte Erwartung *f*
unconditional expression unbedingter Ausdruck *m*
unconditional identity rule Satz *m* von der unbedingten Identität
unconditioned response unbedingtes (bedingungsfreies) Verhalten *n* *{normales Verhalten bei einem Reiz, z.B. Angst bei einem Brand}*
unconditioned stimulus unbedingter Reiz (Stimulus) *m* *{normaler Reiz, der ein unbedingtes Verhalten hervorruft}*
unconstrained finite state unbeschränkter (freier) Endzustand *m*
unconstrained maximization problem Maximierungsproblem *n* ohne Randbedingungen
unconstrained minimization problem Minimierungsproblem *n* ohne Randbedingungen
uncorrectable error unkorrigierbarer (nichtkorrigierbarer) Fehler *m*
undecidability Nichtentscheidbarkeit *f*
undecidable unentscheidbar, nicht entscheidbar
unessential game unwesentliches Spiel *n*
unification Unifikation *f*, Unifizierung *f*, *{Prozedur für Fallbetrachtungen, die Substitutionen für Variabe sucht, mit denen zwei Atome identisch gemacht werden können}*
union axiom Vereinigungsmengenaxiom *n*
unique eindeutig *{Mathematik}*
uniqueness Eindeutigkeit *f* *{Mathematik}*
uniqueness operator Eindeutigkeitsoperator *m*
uniqueness theorem Eindeutigkeitssatz *m*
unity-square game Spiel *n* über dem Einheitsquadrat
universal algorithm universeller Algorithmus *m*
universal function universelle Funktion *f*
universal logic module universeller Logikbaustein *m*
universal quantifier Allquantifikator *m*, Allquantor *m*, Alloperator *m*, Allzeichen *n*, universeller Quantikator *m* *{Logik}*
universal sentence Universalaussage *f* *{Logik}*
universal set Allmenge *f* *{Logik}*
universal Turing machine universelle Turing-Maschine *f*

universal validity Allgemeingültigkeit *f* {*Logik*}
universality Universalität *f*, Allgemeingültigkeit *f*
universally valid logisch gültig, prädikatenlogisch allgemeingültig
universally valid form allgemeingültiger Ausdruck *m*
unknown unbekannt; unbestimmt {*Wahrheitswert in der nichtklassischen Semantik*}
unknown quantity unbekannte Größe *f*, Unbekannte *f*
unlink/to Verbindung (Verknüpfung) lösen {*zwischen Dateien oder zwischen Programmen*}
unoccupied place unbesetzter Platz *m* {*Bedienungstheorie*}
unpredictable nicht vorhersagbar
unsolvable problem unlösbares Problem *n*
unsolved conflict ungelöster Konflikt *m*
unstable system instabiles System *n*
unsteadiness Unstetigkeit *f*
unsteady unstetig
unstructured decision nichtstrukturierte Entscheidung *f* {*nicht durch Regeln beschreibbar*}
unsupervised learning unüberwachtes Lernen *n*
untrue falsch {*Logik*}
update/to 1. pflegen, aktualisieren {*z.B. Datei*}; 2. [zeittreu] nachführen {*Modell*}; 3. [zwecks Aktualisierung] korrigieren {*Speicherinhalt*}
updating 1. Aktualisierung *f*; Wartung *f*, Pflege *f* {*Datei*}; Nachführung *f* {*Modell*}

useful information Nutzinformation *f*
user model Benutzermodell *n*

V

valence 1. Stellenzahl *f* {*logische Operation*}; 2. Valenz *f*, Wertigkeit *f* {*z.B. Graph*}
valence of node Valenz *f* des Knotens {*Graph*}
valid [logisch] gültig, zulässig
valid argument gültige Aussagenverbindung *f* {*Logik*}
valid colourable graph zulässig färbbarer Graph *m*
valid expression gültiger Ausdruck *m*
valid in predicate logic *s*. universally valid
valid strategy zulässige Strategie *f*
validation Gültigkeitserklärung *f*, Bestätigung *f*, Beweis *m*
validation parameter Validierungsparameter *m*
validity Gültigkeit *f*; Treffsicherheit *f*, Validität *f* {*Prognose*}
validity check[ing] Gültigkeitsprüfung *f*, Gültigkeitskontrolle *f*
value Wert *m* {*z.B. einer Variablen*}
variable Variable *f* {*Größe, Menge oder Funktion, die erforderlichenfalls innerhalb eines vorgegebenen Wertebereiches jeden vorgegebenen Wert annehmen kann*}
vector Vektor *m* {*geordneter Liste von Zahlen*}
virtual image virtuelles Bild *n* {*projiziertes Bild*}
vision Bilderkennung *f*

vision module Bilderkennungsmodul *m*
voice recognition Stimmenerkennung *f*
von Neumann architecture von Neumann-Architektur *f*, sequentielle Computerarchitektur *f*

W

weight Gewichtsfaktor *m*, Gewicht *n* *{Wert der Signale am Eingang eines Neurons}*
weight matrix Gewichtsmatrix *f*
weighted matrix gewichtete Matrix *f*
window function Fensterfunktion *f*
work cell Bearbeitungszelle *f* *{computergesteuerte Maschinengruppe für gemeinsame Arbeitsaufgabe}*
work zone Arbeitsbereich *m* *{Roboter}*
working memory Arbeitsspeicher *m*, Kurzzeitgedächtnis *n*, dynamische Wissensbank *f*, Faktendatenbank *f*
world knowledge Weltwissen *n* *{Wissen über einen für eine Problemlösung interessierenden Teil der Welt}*
world model Weltmodell *n* *{Darstellung der aktuellen Situation des für eine Problemlösung interessierenden Teiles der Welt}*

Y

yaw/to gieren *{Roboterbewegung}*
yes-or-no answer Ja/Nein-Antwort *m*

Z

zero memory source Quelle *f* ohne Gedächtnis

Deutsch/Englisch

A

abarbeiten run/to *{program}*
abbilden map/to *{mathematics}*
Abbilden *n* mapping
Abbildung *f* figure; image; map, mapping
Abbildungsbereich *m* feature map *{areas of the brain which are directly connected to sensory parts of the body}*
Abbildungsinversion *f* map inversion
abbrechen abort/to *{e.g. program}*
Abbruch *m* [abnormal] termination
Abbruchfehler *m* truncation error
Abelsche Gruppe *f* Abelian group
abfahren playback/to *{robot}*
Abfertigung *f* serving *{queueing system}*
Abfertigung *f* **mit Prioritäten** priority serving *{queueing system}*
Abfrage *f* inquiry, interrogation, query; retrieval *{data base}*; consultation *{expert system}*
abfragen inquire/to, interrogate/to; retrieve/to *{data base}*; consult/to *{expert system}*
Abfragesprache *f* query language
Abfragesystem *n* interrogation system
Abfühlen *n* sensing
abgegrenzt definite
Abgleich *m* adjustment
Abgleichproblem *n* adjustment (trimming) problem
abhängen depend/to
Abhängigkeit *f* dependence, dependency
Abhängigkeit *f* **vom System** system dependency
abhängigkeitsgesteuert dependency-directed
Abhängigkeitsgraph *m* dependence graph
Abklingen *n* attenuation, decay *{signal}*
Abkürzungsschema *n* abbreviation scheme
Ablauf *m* flow, progress, sequence
ablaufen progress/to *{e.g. program}*
Ablaufgraph *m* sequence graph
Ablaufplan *m* schedule
Ableitbarkeitsrelation *f* relationship of derivability
abriegeln interlock/to *{e.g. signal flow}*
Abrundung *f* truncation
abschätzen evaluate/to
Abschneidefehler *m* truncation error
Abschwächung *f* attenuation
absolut absolute
absolut autonomes System *n* absolutely autonomous system
absolute Adresse *f* absolute address
absolute Adressierung *f* absolute addressing
absolute Codierung *f* absolute coding
absoluter Assembler *m* absolute assembler
absoluter Code *m* absolute code
absoluter Extremwert *m* absolute extreme value
absoluter Vektor *m* absolute vector
absolutes Extremum *n* absolute extreme value
absolutes Kriterium *n* absolute criterion
absolutes Maximum *n* global maximum
absolutes Minimum *n* global minimum

abspielen

absolutes Optimum *n* absolute (global) optimum
abspielen playback/to
Abstiegsmethode *f* descent method
Abstiegsweg *m* descent path
abstrakt abstract
  abstrakte Automatentheorie *f* abstract automata theory
  abstrakte Maschine *f* abstract machine
  abstrakte Syntax *f* abstract syntax
  abstrakte Synthese *f* **von Automaten** abstract synthesis of automata
  abstrakter Automat *m* abstract automaton, abstract machine
  abstrakter Graph *m* abstract graph
  abstrakter Operator *m* abstract operator
  abstrakter Raum *m* abstract space
  abstraktes Modell *n* abstract (symbolic) model
Abstraktion *f* abstraction
Abstraktionsgrad *m* degree of abstraction
Abstraktionsniveau *n* abstraction level
Abstraktionsprozeß *m* abstraction process
Abstraktionsstufe *f* abstraction level
Absturz *m* [abnormal] termination
Abtasten *n* sensing
Abtastung *f* sampling
Abtastzeit *f* action period (phase, time) *{systems theory}*
Abtrennungsregel *f* separation rule *{predicate logic}*
Abwandlung *f* modification
Abwärtssteuerung *f* subcontrol *{learning automaton}*
Abweichung *f* error

124

abzählbare Menge *f* denumerable set
Abzweig *m* branch
Abzweigstruktur *f* branch structure
Abzweigung *f* branching
Achsenkreuz *n* coordinate system
Adaptation *f* adaptation, self-adaptation
Adaptationsalgorithmus *m* adaptation algorithm
Adaptationsgeschwindigkeit *f* adaptation speed
Adaptationsgesetz *n* adaptation law
adaptieren adapt/to
adaptionsfähig adaptive
Adaptionsfähigkeit *f* adaptability *{ability of a network to modify itself in response to changing conditions}*
Adaptionsschleife *f* adaptive loop
Adaptionssystem *n* adaptive system
adaptiv adaptive
  adaptive Abtastung *f* adaptive sampling
  adaptive Bahnführung *f* adaptive routing
  adaptive Fertigungsstraße *f* adaptive process line
  adaptive Kanalzuweisung *f* adaptive channel allocation
  adaptive lineare Schaltung *f* adaptive linear network
  adaptive Logik *f* adaptive logic
  adaptive Regelung *f* **mit Bezugsmodell** model-reference adaptive control
  adaptive Resonanz *f* adaptive resonance
  adaptiver Automat *m* adaptive automaton
  adaptiver Korrektor *m* adaptive corrector

adaptiver lokaler Optimierungsalgorithmus *m* adaptive local optimization algorithm
adaptiver Regler *m* adaptive controller
adaptives digitales Element *n* adaptive digital element
adaptives Filter *n* adaptive filter *{filtering process changes automatically as the characteristics of the incoming signals change}*
adaptives Lernsystem *n* adaptive learning system
adaptives Prozeßmodell *n* adaptive process model
adaptives Schwellenwertelement *n* adaptive threshold element, adaptive linear network
adaptives System *n* adaptive [control] system
Adaptivkreis *m* adaptive loop
Adaptivkreisverstärkung *f* adaptive-loop gain
Adäquatheit *f* adequacy
additiv additive
Additivität *f* additivity
adjazent adjacent
Adjazenz *f* adjacency
Adjazenzmatrix *f* adjacency matrix *{graph theory}*
adjungiert adjoint
  adjungierte Funktion *f* adjoint function
  adjungierte Matrix *f* adjoint matrix
  adjungierte Systemgleichung *f* adjoint system equation
  adjungierte Variable *f* adjoint variable
  adjungierter Ausdruck *m* adjoint expression

adjungierter Operator *m* adjoint operator
adjungiertes System *n* adjoint system
Adjunktion *f* adjunction
adressierbar addressable
adressierbare horizontale Position *f* addressable horizontal position
adressierbare vertikale Position *f* addressable vertical position
adressierbarer Punkt *f* addressable point
Adressierung *f* addressing
Adreßbuch *n* **mit Baumstruktur** tree-structured directory
aeq aeq *{truth function, IF AND ONLY IF}*
Agenda *f* list *{ordered list of actions}*
Aggregation *f* aggregation
Aggregatmodellierung *f* **großer Systeme** aggregate model[l]ing
ähnlich similar
Ähnlichkeit *f* similarity
Ähnlichkeit *f* **im Großen** large-scale similarity
Ähnlichkeit *f* **im Kleinen** small-scale similarity
Ähnlichkeitserkennung *f* similarity recognition
Ähnlichkeitsfunktion *f* similarity function
Ähnlichkeitskriterium *n* similarity criterion
Ähnlichkeitsmaß *n* similarity measure
Ähnlichkeitsmatrix *f* similarity matrix
Ähnlichkeitsprinzip *n* principle of similarity, similarity principle

Ähnlichkeitstheorie

Ähnlichkeitstheorie *f* similarity theory, theory of similarity
Akkretionsnetz *n* accretive network *{gathers a clear date set out of a noisy set}*
Akkumulator *m* accumulator
akkumulieren accumulate/to
akkumulierter Fehler *m* accumulated error
Aktion *f* action
Aktionsliste *f* action list *{ordered list of actions}*
aktionsorientierte Darstellung *f* procedural representation
Aktionspfad *m* action path
Aktionspotential *n* action potential *{potential which characterizes one of the two states excited or unexcited when the membrane potential crossed over the threshold}*
Aktionsteil *m* [einer Regel] right-hand side, action part
aktiv active
   aktive Signalebene *f* assertion level
   aktiver Wert *m* default value
   aktives Netzwerk *n* active network
   aktives Signal *n* asserted signal
aktivieren activate/to
aktiviert active
Aktivierung *f* activation
Aktivierungsfunktion *f* activation function *{specifies what the neuron is to do with the signal after the weights have had their effects}*
Aktivierungsmechanismus *m* activation mechanism
Aktivierungsniveau *n* activation level
Aktivierungswert *m* activation value *{sum of weighted inputs of a neuron at a given time}*
Aktivierungszustand *m* activation state *{state of a single neuron or a network}*
Aktivität *f* activity
Aktivitätsdauer *f* activity duration
Aktivitätsnummer *f* activity number
aktivitätsorientierter Netzplan *m* activity-on-node network, activity-oriented network
Aktivitätstermin *m* activity time
aktualisieren update/to *{e.g. file}*
Aktualisierung *f* updating
aktuell actual
   aktueller Parameter *m* actual parameter
   aktueller Wert *m* actual value
akustische Ausgabe *f* acoustic output
akustische Eingabe *f* acoustic input
akustischer Dialog *m* acoustic dialogue *{man-machine system}*
akzeptierbar acceptable
akzeptierbare Menge *f* semi-calculable set
Alephhypothese *f* aleph hypothesis *{generalization of the continuum problem}*
Algebra *f* algebra
   Algebra *f* **der Algorithmen** algebra of algorithms
   Algebra *f* **der Logik** algebra of logic, logic algebra
algebraisch algebraic
   algebraische Automatentheorie *f* algebraic automata theory
   algebraische Notation *f* algebraic notation
   algebraische Struktur *f* algebraic structure

algebraische Topologie *f* algebraic topology
algebraischer Ausdruck *m* algebraic expression
algebraisches Stabilitätskriterium *n* algebraic stability criterion
Algorithmen- und Programmbibliothek *f* algorithm and program library
Algorithmenbegriff *m* algorithmic concept
Algorithmenschema *n* algorithm scheme
Algorithmentheorie *f* algorithm theory
Algorithmierung *f* algorithmization
Algorithmierung *f* **kreativer Prozesse** algorithmization of creative processes
Algorithmierung *f* **von Produktionsprozessen** algorithmization of production processes
algorithmisch algorithmic
algorithmische Logik *f* algorithmic logic
Algorithmus *m* algorithm *{procedure for problem solving in a finite number of steps}*
Algorithmus *m* **für verborgene Oberflächen** hiding algorithm *{graphic representation}*
Aliasname *m* alias name
Alles-oder-Nichts-Gesetz *n* all-or-none law
allgemein general
  allgemeine Erlang-Verteilung *f* general Erlang distribution
  allgemeine mengentheoretische Topologie *f* general topology
  allgemeine Systemtheorie *f* general systems theory
  allgemeiner Problemlöser *m* general problem solver
allgemeingültig universally valid
allgemeingültiger Ausdruck *m* universally valid form
Allgemeingültigkeit *f* generality, universality, universal validity
Allmenge *f* universal set *{logic}*
Alloperator *m* universal quantifier *{logic}*
Allquantifikator *m* universal quantifier *{logic}*
Allquantor *m* universal quantifier *{logic}*
Alltagssituation *f* everyday situation
Allzeichen *n* universal quantifier *{logic}*
Allzweckroboter *m* general-purpose robot
Alphabet *n* alphabet
alphabetisch alphabetic
  alphabetischer Code *m* alphabetic code
  alphabetischer Zeichensatz *m* alphabetic character set
  alphabetisches Zeichen *n* alphabetic character
Alphabetoperator *m* alphabetic operator
Alphabetzeichen *n* alphabetic character
Alphanummer *f* alpha number
Alphasprache *f* alpha language
Alphasystem *n* alpha system
Alphazeichen *n* character
Alternative *f* inclusive OR function
Alternativhypothese *f* alternative hypothesis

Alternator *m* alternator *{system theory}*
alternierende Reihe *f* alternating series
analoge Information *f* analogous information
analoge Simulierung *f* analogous simulation
Analogie *f* analogy
Analogiemodell *n* analog[ue] model
Analogieprinzip *n* analog[ue] principle
Analogierelation *f* analog[ue] relation
Analogieschluß *m* analog[ue] conclusion
Analoginformation *f* analog[ue] information
Analogsimulierung *f* analog[ue] simulation
Analyse *f* **der entgangenen Gewinne** optimality criterion of Savage
Analyse *f* **von Automaten** analysis of automata
Analysenproblem *n* analysis problem
Analyseverfahren *n* analysis technique
analysieren analyze/to
analytische Beziehung *f* analytical relation
analytische Fortsetzung *f* analytical continuation
analytische Funktion *f* analytical function
analytische Lösung *f* analytical solution
analytische Umformung *f* analytical transformation
analytisches Prozeßmodell *n* analytical process model
analytisches Sprachmodell *n* analytical language model
Andersname *m* alias name
Änderung *f* change, modification
aneinanderreihen chain/to
anfänglich initial
Anfangsbedingung *f* initial condition
Anfangsereignis *n* starting (beginning) event *{graph}*
Anfangsfunktion *f* starting function
Anfangsknoten apex (root) node *{tree structure}*
Anfangskonzept *n* initial conception
Anfangslösung *f* starting solution *{e.g. linear optimization}*
Anfangsverhalten *n* initial behaviour
Anfangswert *m* minimum value
Anfangswertproblem *n* initial-value problem
Anfangszustand *m* initial condition
Anfrage *f* inquiry, question
angenäherte Lösung *f* approximate solution
angenähertes Modell *n* approximation model
angesammelt accumulated
angewandte Linguistik *f* applied linguistics
Angriff *m* attack
anhalten pause/to
Animation *f* animation
Ankunft *f* arrival *{queueing theory}*
Ankunftprozeß *m* arrival process
Ankunftsrate *f* arrival rate
Ankunftsreihenfolge *f* arrival sequence
Ankunftsstrom *m* arrival stream
Ankunftszeit *f* arrival time

anlagenabhängig system-dependent
Anlagenausfall *m* system failure
Anlagenzuverlässigkeit *f* system reliability
Anmerkung *f* remark
annähernd approximate, approximating
Annäherung *f* approximation, approach
Annahme *f* assumption, belief
annehmbar acceptable
anordnen order/to, sequence/to
Anordnen *n* **von Objekten nach Klassen gemäß vergegebenen Konzepten** conceptual clustering
Anordnung *f* arrangement, array, configuration, pattern
Anordnungsrelation *f* order relation, relationship of order
anpaßbar adaptable
anpassen adapt/to, match/to
Anpassung *f* adaptation, adaption, matching
anpassungsfähig adaptable, self-adaptive; *{s.a. adaptiv}*
anpassungsfähiger Regelkreis *m* adaptive loop
Anpassungsfähigkeit *f* adaptability; plasticity *{ability of a group of neurons to adapt their function to a different function over time and to take over the functions of a damaged portion of a neuronal network}*
Anpassungsproblem *n* matching problem
anregen activate/to *{e.g. oscillations}*
ansammeln accumulate/to
Ansammlung *f* accumulated error
Ansatz *m* approach *{mathematics}*
Anschlußleitung *f* connecting line

Anschlußproblem *n* matching problem
ansprechen [auf] respond/to
Ansprechen *n* response
Ansprechzeit *f* response time
Ansteuerlogik *f* controller logic
ansteuern control/to
Anstiegsweg *m* ascent path *{gradient method}*
antagonistisches Spiel *n* antagonistic game
Antecedens *f* antecedent *{left side of production rule}*
Anthropotechnik *f* human engineering
Antinomie *f* antinomy
antinomische Menge *f* antinomous set
Antivalenz *f* antivalence, contravalence, exclusive OR, exclusive OR operation, non-equivalence
Antivalenzelement *n* non-equivalence element
Antivalenzoperation *f* non-equivalence operation
Antivalenzschaltung *f* exclusive OR circuit (element, gate, network)
Antizickzackvorkehrung *f* antizigzag measure (provision) *{gradient method}*
Antwort *f* **des Erkennungssystems** response of pattern recognition system
antwortendes System *n* question-answering system
Antwortüberlappung *f* interleave
Anwendung *f* application
Anwendung *f* **der Wahrscheinlichkeitsrechnung bei der Mustererkennung** probability use in pattern recognition

Anwendungsbereich *m* range of use
anzeigen display/to
Approximation *f* approximation, *{s.a.* Näherung*}*
Approximation *f* **durch Iteration** approximation by iteration
Approximation *f* **einer Funktion** approximation of function
Approximationsfehler *m* approximation error
Approximationsfunktion *f* approximation function
Approximationsgebiet *n* approximation region
Approximationsintervall *n* approximation interval
Approximationsmethode *f* approximation method
Approximationsmodell *n* approximation model
Approximationspolynom *n* approximation polynomial
Approximationstheorie *f* approximation theory
Approximationswert *m* approximation value
approximativ approximate
äquiprobabel equiprobable
äquivalent equivalent
Äquivalenz *f* equivalence
Äquivalenz *f* **von Spielen** strategic equivalence
Äquivalenzbeziehung *f* equivalence relation
Äquivalenzelement *n* equivalence element
Äquivalenzgatter *n* equivalence element
Äquivalenzglied *n* equivalence element
Äquivalenzoperation *f* equivalence operation
Äquivalenzprinzip *n* equivalence principle
Äquivalenzrelation *f* equivalence relation
Äquivalenzschaltung *f* equivalence element
arbeiten work/to, run/to
arbeitend active
Arbeitsaufgabe *f* job, task
Arbeitsbereich *m* work zone *{robot}*
Arbeitshilfe *f* job aid
Arbeitsspeicher *m* working memory
Arbeitsspeicher *m* **für die Tafelmethode** blackboard
Arbeitsweise *f* function, mode, principle of operation
Architektur *f* architecture
Argument *n* argument, arg
Argumentation *f* argumentation; argument form, reasoning *{logic}*
Arithmetikeinheit *f* unit arithmetic organ, arithmetic logic unit
arithmetisch arithmetic[al]
arithmetische Funktion *f* arithmetic function
arithmetische Operation *f* arithmetic operation, computing operation
arithmetische Reihe *f* arithmetic series
arithmetische und analytische Hierarchie *f* arithmetical and analytical hierarchy
arithmetischer Elementarausdruck *m* arithmetic element
arithmetischer Operator *m* arithmetic operator
Armpositionierung *f* arm positioning *{robot}*
Array *n* array

Arraylogik *f* array logic
Arrayprozessor *m* array processor
Art *f* mode
Assistenzprogramm *n* assistant program
assoziatives Element *n* associative element
assoziatives Gesetz *n* associative law
assoziatives Speichernetz *n* associative memory network *{a type of network which has the ability to recognize patterns or objects}*
Assoziativgesetz *n* associative law
Assoziativität *f* associativity
Assoziator *m* associator; *{s.a.* assoziatives Speichernetz*}*
astabil astable
asymmetrisch asymmetric[al], nonsymmetric
asymptotisch asymptotic
asymptotische Stabilität *f* asymptotic stability
asymptotischer Endwert *m* asymptotic final value
asymptotisches Verhalten *n* asymptotic behaviour
asynchrone Logik *f* non-symmetric logic
asynchroner Automat *m* asynchronous automaton
asynchrones sequentielles Schaltwerk *n* asynchronous sequential circuit
Atom *n* atom *{data element or indivisible logic expression}*
atomar atomic
atomare Formel *f* atomic formula
atomare Funktion *f* logic elementary function
atomarer Ausdruck *m* atomic expression
atomares System *n* atomic system
Atomausdruck *m* atomic expression
Attraktionsbecken *n* basin of attraction *{area surrounding the energy minimum on the E surface}*
Attribut *n* attribute *{property of an entity, e.g. name, age etc. of a person}*
Attributgrammatik *f* attributive grammar
Aufbau *m* arrangement, structure
Aufbaumethode *f* bottom-up method
aufbereiten condition/to
aufeinanderfolgend consecutive, sequential
Auffindung *f* **von Rechtschreibfehlern** spelling error detection
Aufgabe *f* function, job, problem, task
aufgabenorientierte Sprache *f* object language, task-oriented language
Aufgabenstellung *f* problem, problem formulation
aufgabenunabhängig problem-independent
Aufgabenwarteschlange *f* task queue
aufgelaufener Fehler *m* accumulated error
aufgerufen active *{e.g. file}*
aufgespannter Graph *m* stretched graph
aufgeteilt sliced
aufgeteilter Bildschirm *m* split screen
Auflösung *f* resolution
Auflösungsgenauigkeit *f* accuracy of resolution
Auflösungsvermögen *n* acuity

Aufnahme f record
Aufnahme- und Plaziereinheit f pick-and-place unit *{robot}*
Aufnehmen n perception *{e.g. pattern}*
Aufruf m call, invokation
aufrufen invoke/to
Aufrundung f rounding-up
aufrüstbar expandable
aufspannen stretch/to *{graph}*
aufsteigen ascend/to *{e.g. order}*
Aufstiegsmethode f ascent method
Aufstiegsweg m ascent path
Aufteilen n sharing
Auftrag m job
auftreten occur/to
Aufwärtssteuerung f supercontrol *{learning automaton}*
Aufzeichnung f record *{part of a file}*
augenblicklich current
Ausbildungssoftware f instructional software, teachware
ausdehnbar expandable
Ausdehnung f extension; expansion
Ausdruck m expression, form *{logic}*
Ausfall m des Systems system failure
Ausfall m einer Regelstrecke (Strecke) f system failure
ausführen perform/to *{task}*
ausführen/eine logische Operation perform a logic operation/to
Ausführung[sart] f configuration
Ausführungsform f model
Ausführungshilfe f job (performance) aid
Ausgabeneuron n output neuron *{sends information out to the human or to something else}*
Ausgabeprozeß m [queueing]
output process
Ausgabewarteschlange f output queue
Ausgangsfunktion f starting function
Ausgangslogik f output logic
Ausgangslösung f starting solution
Ausgangswert m original value
ausgeben output/to
ausgeglichener Baum m balanced tree
ausgeklügelt sophisticated
ausgeprägtes Extremum n pronounced extremum
ausgeprägtes Maximum n pronounced maximum
ausgeprägtes Minimum n pronounced minimum
ausgeprägtes Optimum n pronounced optimum
ausgeschlossener Widerspruch m excluded contradiction
Ausgleich m adjustment; self-regulation
Ausgleich m bedingter Beobachtungen adjustment of conditional observations
ausgleichen balance/to; control/to
Ausgleichsrechnung f adjustment-of-data calculus
Auslastungsregelung f adaptive control with constraints *{numerical control}*
auslösen activate/to; actuate/to *{directly by man}*; fire/to
Auslösung f actuation *{directly by man}*; activation *{e.g. relais}*
Ausnahme f exception
ausprobieren try/to
Aussage f declarative sentence, statement, proposition *{logic}*

Aussageform *f* sentential (propositional) form *{predicate logic}*
Aussagenfunktion *f* propositional function
Aussagengröße *f* Boolean variable
Aussagenkalkül *m* propositional calculus, sentential calculus
Aussagenlogik *f* assertion logic, propositional logic
aussagenlogisch Boolean
aussagenlogisch unzerlegbar indecomposable in propositional logic
aussagenlogisch wahr tautological
aussagenlogisch zerlegbar decomposable in propositional logic
aussagenlogische Identität *f* propositional-logic identity
aussagenlogische Konstante *f* propositional-logic functor
aussagenlogischer Ausdruck *m* propositional-logic expression
aussagenlogischer Funktor (Junktor) *m* propositional-logic functor
Aussagenvariable *f* Boolean (propositional) variable
Aussagenverbindung *f* argument
Aussagenverknüpfung *f* propositional connection
ausscheidende Induktion *f* eliminating induction
ausschließende ODER-Operation *f* exclusive OR operation, non-equivalence *{s.a. exklusive OR}*, antivalence
ausschließende ODER-Schaltung *f* exclusive OR circuit (element, gate, network)
ausschließendes ODER-Gatter (ODER-Tor) *n* exclusive OR circuit (element, gate, network)

ausschließlich exclusive
Ausschließung *f* exclusion
Außenwelt *f* outside world
äußere Logik *f* external logic
äußere Prozeßanalyse (Systemanalyse) *f* terminal analysis [of a system]
äußerer Input *m* external input *{system theory}*
äußeres Modell *n* external model
Austausch substitution, interchange
Austauschbarkeit *f* exchangeability
austauschen durch (mit) substitute by/to
Auswahlaxiom *n* axiom of selection
auswählen select/to
Auswahllogik *f* selection logic
Auswahlmenge *f* selective set
Auswahlprinzip *n* axiom of choice *{set theory}*
Auswechselbarkeit *f* exchangeability
ausweichen balk/to *{queueing theory}*
auswerten analyze/to, evaluate/to, interpret/to
Auswertung *f* evaluation
Auswertungsfunktion *f* evaluation function
Auswertungssystem *n* evaluation system
Auszahlungsfunktion *f* profit (gain) function *{theory of games}*
Auszahlungsmatrix *f* paying-off matrix, profit (gain) matrix *{theory of games}*
Authentifikation *f* authentification
Authentifizierung *f* authentification
Authentisierung *f* **von Nachrichten** authentification of messages
Authentizität *f* authenticity

autoassoziatives Speichernetz *n* autoassociative memory network *{inputting incomplete information causes the network to output the complete memory}*
Autokorrelation *f* autocorrelation
Autokorrelationsanalyse *f* autocorrelation analysis
Autokorrelationsfunktion *f* autocorrelation function
Autokorrelationsmatrix *f* autocorrelation matrix
Autokovarianzfunktion *f* autocovariance function
Automat *m* automaton, machine
Automat *m* **mit endlichem Speicher** finite-memory automaton
Automat *m* **mit gespeicherter Logik** stored-logic computer
Automat *m* **mit [nur] einem Zustand** single-state automation
Automat *m* **ohne Gedächtnis** memoryless automaton, automaton without memory
Automatendiagnose *f* computer diagnosis
Automatennetz *n* automata network
Automatenoperator *m* automata operator
Automatensprache *f* automata language
Automatentabelle *f* automata table, state table
automatentheoretische Mikroprogrammdarstellung *f* automata-theoretical microprogram representation
Automatentheorie *f* automata theory, theory of automata
automatisch automatic, self-regulating

automatisch zurückrückend autodecrement
automatische Codierung *f* automatic coding
automatische Diagnostik *f* automatic diagnostics
automatische Fehlerauffindung *f* automatic error detection
automatische Indizierung *f* automatic indexing
automatische Klassifikation *f* automatic classification
automatische logische Beweisführung *f* automatic reasoning
automatische Sprachanalyse *f* automatic language analysis, machine analysis
automatische Spracherkennung *f* automatic speech recognition
automatische syntaktische Analyse *f* automatic syntactical analysis
automatische Übersetzung *f* automatic language translation, automatic translation, machine translation
automatisches Beweisen *n* **von Theoremen** automatic theorem proofing
automatisches Fehlerkorrektursystem *n* automatic error-correcting system
automatisches logisches Schließen *n* automatic reasoning
automatisiert automated
automatisierter Unterricht *m* automated teaching
automatisiertes Beweisen *n* **von Theoremen** automated theorem proofing (reasoning)
automatisiertes Managementsystem *n* automated management system

automatisiertes System *n* **der Arbeitsorganisation** automated work organization system
automatisiertes System *n* **der Projektierung** automated system of project design
Automatisierung *f* automati[zati]on
Automatisierung *f* **der medizinischen Diagnostik** automation of medical diagnostics
Automatisierung *f* **der Therapie** automation of therapy
Automorphismus *m* automorphism
autonom autonomous
 autonomer Automat *m* autonomous automaton
 autonomes Rechnersystem *n* autonomous computer system
 autonomes System *m* autonomous system
Autorepeat-Funktion *f* autorepeat function, cycling
Axiom *n* axiom
axiomatisch axiomatic
 axiomatische Semantik *f* axiomatic semantics
 axiomatische Wahrscheinlichkeitsrechnung *f* axiomatic calculus of probability
Axiomensystem *n* axiom system
Axon *n* axon {extension of a biological neuron which carries the output signal away to other neurons}

B

B-Baum *m* B-tree
Back-up-System *n* back-up system
Bahn *f* trajectory; path {numerical control}
Bahnführung *f* routing {robot}
Balanceproblem *n* pole balancing problem
Banach-Raum *m* Banach space
Barrieremethode *f* interior penalty method
Basismatrix *f* basic matrix
Bauelemente *npl* hardware
Baugruppe *f* constructional subassembly, subset
Baum *m* tree {e.g. decision tree}
baumähnlich tree-like
Baumdarstellung *f* tree representation
Baumstruktur *f* tree structure
Baumsuche *f* tree search
Baumsuchverfahren *n* tree-search technique
Baustein *m* module
Bayes-Entscheidungsregel *f* Bayes' decision rule
Bayes-Klassifikator *m* Bayes' classifier {theory of games}
Bayes-Kriterium *n* Bayes' criterion
Bayes-Schätzung *f* Bayes' estimation
Bayessches Lehrverfahren *n* Bayes' teaching
Bearbeitungszelle *f* work cell
Bedarf *m* demand
Bedarfsvorhersage *f* demand predicition
bedecken cover/to
Bedeutung *f* signification; meaning {e.g. of a signal}

Bedeutungsanalyse *f* content analysis
bedeutungsbezogen semantic, related to meaning
bedeutungslos meaningless
bedienen actuate/to, control/to
Bediener *m* server
Bediengraph *m* queueing network
Bedienprozeß *m* serving process
bediente Forderung *f* served demand (requirement)
Bedienung *f* actuation
  Bedienung *f* **durch Spracheingabe** oral control
  Bedienung *f* **mit zufälliger Auswahl** service in random order
Bedienungsanforderung *f* service request
Bedienungsapparat *m* server
Bedienungsdauer *f* detention time
Bedienungsintensität *f* serving intensity
Bedienungsnetz *n* queueing network
Bedienungsprozeß *m* service (serving) process
  Bedienungsprozeß *m* **mit einem Bedienungsapparat** single-server queueing process
Bedienungssystem *n* queueing system
  Bedienungssystem *n* **mit absoluter Priorität mit Verlust** preemptive loss-queueing system
  Bedienungssystem *n* **mit begrenzter Verweildauer** queueing system with limited residence period
  Bedienungssystem *n* **mit begrenzter Wartezeit** queueing system with limited latency time
  Bedienungssystem *n* **mit dynamischer Priorität** dynamic-priority queueing system
  Bedienungssystem *n* **mit Zeitbegrenzungen** queueing system with time limits
Bedienungstheorie *f* congestion theory, queueing theory
bedingt conditional, ONLY THAN IF
bedingte Beobachtung *f* conditional observation
bedingte Entropie *f* average conditional information content
bedingte Erwartung *f* conditional expectation
bedingte Hemmung *f* conditioned inhibition
bedingte Implikation *f* conditional implication
bedingte Informationsentropie *f* average conditional information content
bedingte Stabilität *f* conditional stability
bedingte Verknüpfung *f* conditional connection
bedingte Wahrscheinlichkeit *f* conditional probability
bedingter Code *m* conditional code
bedingter Erwartungswert *m* conditional expectation value
bedingter Informationsgehalt *m* conditional information content
bedingter Reflex *m* conditional reflex
bedingter Reiz *m* conditioned stimulus
bedingtes Ansprechen *n* conditioned response
bedingtes Ereignis *n* conditional event

bedingtes Verhalten *n* conditioned response
Bedingung *f* condition
Bedingung *f* **für das zeitliche Gleichgewicht** steady-state condition
Bedingungseingang *m* condition input
Bedingungsteil *m* condition part *{logic condition for the decision}*
Bedingungsteil *m* **einer Regel** left-hand side [of a rule]
Beeinflussung *f* influence, action
beenden/[einen Prozeß] abort [a process]/to
Befehl *m* instruction, order
befehlbefolgender Automat *m* instruction-obeying machine
Befehlsaufbau *m* instruction structure
Befehlsliste *f* instruction list (set)
Befehlsmenge *f* instruction set
Befehlssatz *m* instruction set
Befehlsvorrat *m* instruction set
Befestigungsmittel *npl* hardware
befriedigen satisfy/to *{demand}*
Begleittext *m* context
begrenzen bound/to; limit/to
Begrenzer *m* hard limiter
begrenzt constrained
begrenzter Bereich *m* bounded domain
Begrenzung *f* saturation *{link}*
Begriff *m* concept *{generally}*
begrifflich conceptual
begriffliche Einheit *f* entity
begrifflicher Operator *m* *s.* abstrakter Operator
Begriffsunterscheidung *f* term discrimination
Begründung *f* explanation

Behälterpackungsproblem *n* bin packing problem
Behauptung *f* assertion, sentence *{logic}*
Beispiel *n* example, instance, instantiation
beispielgesteuertes System *n* example-driven system, induction
Beladeproblem *n* cargo-loading problem
Belegung *f* assignment; covering *{logic}*
belehren instruct/to
Belehrung *f* **in der Mustererkennung** teaching in pattern recognition
Bellmansches Optimalitätskriterium *n* optimality criterion of Bellman
benachbart adjacent
benachbarte Knotenpunkte *mpl* neighboured nodes
Benutzermodell *n* user model
Benutzeroberfläche *f* front end
Beobachter *m* observer
Beobachtung *f* observation
Beratung *f* consultation
Beratungsparadigma *n* consultation paradigm
Berechenbarkeit *f* computability
berechnen/den Wert eines [logischen] Ausdruckes evaluate an expression/to
berechnet computed
Bereich *m* area, domain, range
bereitstellen acquire/to
Berichtigung *f* adjustment
Berührungsabtastung *f* tactile sensing
Berührungssensor *m* touch sensor
beschaffen acquire/to

Beschaffungszeit *f* obtainment time
beschleunigte Gradientenmethode *f* accelerated gradient method
Beschleunigungsalgorithmus *m* acceleration algorithm
Beschlußkästchen *n* decision box
Beschneiden *n* pruning *{decision tree}*
beschränken restrict/to
beschränkte Suche *f* constraint search
Beschränkung *f* restriction
beschreibbar describable
Beschreibbarkeit *f* describability
beschreiben describe/to, abstract/to
beschreibend descriptive
beschreibende Sprache *f* descriptive (declarative) language
beschreibendes Modell *n* descriptive model
Beschreibung *f* description
Beschreibung *f* **des Erkennungsobjekts** description of the recognition object
Beschreibungsebene *f* description level
Beschreibungssprache *f* description language
Beschriftung *f* marking
Beseitigung *f* **von Fehlern** error debugging
besetzter Platz *m* occupied place *{queueing theory}*
Besetztperiode *f* busy period *{queueing theory}*
Besetztsein *n* congestion
besorgen acquire/to
beständig stable
Bestandteil *m* constituent [part], part
Bestapproximation *f* proximum

bestätigen acknowledge/to, handshake/to
Bestätigung *f* acknowledgement, confirmation, validation
bestellen order/to
Bestellung *f* order
bestimmen define/to
bestimmen/einen Extremwert obtain an extremum/to
bestimmt definite
bestimmtes Spiel *n* definite game
Bestimmung *f* determination
Bestimmungsgleichung *f* defining equation
Bestimmungsgröße *f* parameter
Bestimmungsort *m* destination [location]
betätigen actuate/to, control/to
Betätigen *n* actuating
betätigt actuated
Betätigung *f* actuation
Betätigungselement *n* actuating member
Betätigungsgröße *f* actuating quantity
Betätigungsorgan *n* [end] effector
Betrag *m* amount *{mathematics}*
Betrag *m* **des Vektors** absolute vector
betriebliche Verfahrensforschung *f* operations research
Betriebsanalysenmeßeinrichtung *f* process analyzer
Betriebsart *f* mode
betriebsbereit active
Betriebsführung *f* management
Betriebsprinzip *n* **einer Computerarchitektur** operational principle of a computer architecture
Betriebsweise *f* function, mode
beurteilbar judgeable
beurteilen judge/to

Beurteilungskriterium *n* criterion of judging
bewegen move/to *{cursor}*
Bewegungsablauf *m* sequence of moves
Bewegungsbahn *f* trajectory
Bewegungsfolge *f* sequence of moves
Beweis *m* proof, validation
 Beweis *m* **durch Rekursion** proof by recursion
beweisbar deductive
Beweisbaum *m* proof tree
beweisen prove/to
Beweisen *n* [theorem] proofing
Beweisführung *f* demonstration, proofing, reasoning
Beweistheorie *f* proof theory
bewerten evaluate/to, interpret/to
Bewerter *m* decision element, evaluator
Bewertungsfunktion *f* evaluation function
Bewertungsgröße *f* rating variable
Bewertungssystem *n* evaluation system, rating system
Bewertungsvariable *f* rating variable
bewirken accomplish/to, effect/to; act/to
Bezeichner *m* identifier, name
Bezeichnung *f* signification
beziehen relate/to
Beziehung *f* relation; association
beziehungslos [zueinander] orthogonal
Beziehungstabelle *f* relation table
Bezugslinie *f* datum line
Bezugsmodell *n* reference model
Bezugsmodellsystem *n* model-reference system

Bezugsobjekt *n* reference object
bibliographische Suche *f* bibliographical retrieval
bichromatischer Graph *m* paired graph
bidirectionales assoziatives Netzwerk *n* bidirectional associative network *{generalization of the Hopfield associative network}*
bijektive Funktion *f* bijective function *{set theory}*
bijektiver Homomorphismus *m* bijective homomorphism, isomorphism
Bild *n* figure, map, image
Bildabtaster *m* pattern scanner
Bildabtastung *f* **mit Tse-Computer** tse detection
Bildauswertegerät *n* pattern interpreter
Bilderfassung *f* image acquisition
Bilderkennung *f* image (pattern) recognition, vision
 Bilderkennung *f* **mit Computer** computer (computational) vision
Bilderkennungsmodul *m* vision module
Bildkompression *f* image compression
bildlich darstellen represent pictorially/to
Bildpositionierung *f* pattern positioning
Bildschirmspeicher *m* smart terminal
Bildsynthese *f* image synthesis
Bildungsregel *f* forming rule *{logic synthax}*
Bildverarbeitung *f* pattern processing

Bildverschlechterung

Bildverschlechterung *f* image degradation
Bildverständnis *n* image understanding, computer vision
Bildwiederauffindung *f* pattern retrieval
Bimatrixdarstellung *f* bimatrix representation
Bimatrixspiel *n* two-matrices game
binär binary *{s.a. zweiwertig}*
binäre Entscheidung *f* binary decision
binäre Information *f* binary information
binäre Informationsquelle *f* binary source
binäre Logik *f* binary logic
binäre logische Funktion *f* binary logic function, switching function
binäre Relation *f* binary relation
binäre Schaltfunktion *f* binary logic function
binäre Struktur *f* binary structure
binärer Baum *m* binary tree
binäres Suchen *n* binary search, dichotomizing search
binäres System *n* binary system
binäres Verknüpfungsglied *n* binary combinational element
Binärentscheidung *f* binary decision
Binärmatrix *f* binary matrix
Binärmuster *n* binary pattern, bit configuration
Binärquelle *f* binary source
Binärsignal *n* **in negativer Logik** negative-true binary signal
Binärsuche *f* binary search
Binärsystem *n* binary system
Binärvariable *f* binary variable
Binärzeichen *n* binary character
Bindeglied *n* link

Bindung *f* link
Bindungsstärke *f* connection strength
Binominalreihenentwicklung *f* binominal series expansion
biologische Abbildung *f* biological mapping *{areas of the brain corresponding to body parts and sensory functions}*
biologisches Neuron *n* biological neuron
bipolare Logik *f* bipolar logic
bisektionelles Suchen *n* dichotomizing search
bistabil bistable
bistochastische Matrix *f* bistochastic matrix
Bisubjunktion *f* propositional equivalence
Bitfolge *f* bit sequence
Bitkette *f* bit string
Bitkombination *f* bit combination
Bitmuster *n* bit configuration
bitparallel bit-parallel
bivalent binary
Black-box-Methode *f* black-box method
Blatt *n* terminal node *{search tree}*; leaf *{phase plane}*
Blattknoten *m* leaf node
blinde Suche *f* blind search
Block *m* block; chunk
Blockdiagrammsymbol *n* "Entscheidung" decision box
blockieren block/to, interlock/to
Blockierung *f* blocking *{special case of overshadowing}*
Blockstruktur *f* block structure
Bocksprungstruktur *f* leapfrog structure
Bogen *m* arc, directed edge *{graph}*

Bogenbewertung *f* arc evaluation
Boltzmann-Maschine *f* Boltzmann machine *{network similar to the Hopkins associator but based on the matching of probabilities between the network and the environment}*
Bondgraph *m* bond graph
Bondstelle *f* bond
Boolesche Algebra *f* Boolean algebra
Boolesche Funktion *f* Boolean function
Boolesche Matrix *f* Boolean matrix
Boolesche Operation *f* Boolean operation
Boolesche Operationstafel *f* Boolean operation table, truth table
Boolesche Struktur *f* Boolean structure
Boolesche Variable *f* Boolean variable
Boolesche Verknüpfung *f* Boolean combination
Boolescher Ausdruck *m* Boolean expression
Boolescher Operator *m* Boolean operator
Boolescher Ring *m* Boolean ring
Boolescher Verband *m* Boolean structure
Boolesches Produkt *n* Boolean product
Borelsches Ereignisfeld *n* Borel field of events
Bottom-up-Methode *f* bottom-up approach *{problem-solving begins with simplest tasks and moves to increasingly complex tasks}*
Bottom-up-Strategie *f* bottom-up strategy
brauchbar feasible
brauchbare Richtung *f* feasible direction *{gradient method}*
Breitensuche *f* breadth-first search
Brettschaltung *f* developmental model
Brückenzweig *m* branch
Buchstabe *m* character, letter
Buchstabennummer *f* alpha number

C

Call *m* call
Cauchysches Problem *n* initial-value problem
Cerebralcortex *m* cortex *{the outer portion of a structure}*
Chaos *n* chaos, disorder
chemischer Rezeptor *m* chemical receptor
chemischer Transmitter *m* chemical transmitter
chronologisch chronological
Cluster *n* cluster *{simplest kind of concepts made of symbols, e.g. columns, rows, modules etc.}*
Code *m* Code
Codewort *n* keyword
Codierbarkeit *f* codability
Codierung *f* coding, conversion; encryption
Computer *m* computer, computing automaton, computing machine, machine
Computer *m* **der dritten Generation** third-generation computer
Computer *m* **der fünften Generation** fifth-generation computer

computerabhängig 142

Computer *m* **für Expertensysteme** knowledge machine
Computer *m* **mit starrer Befehlsfolge** consecutive-sequence computer
computerabhängig computer-dependent
Computerarchitektur *f* computer architecture
Computercode *m* machine (absolute) code
Computerdiagnose *f* computer diagnosis
Computergeneration *f* computer generation
computergestützte Fragenbeantwortung *f* computerized question answering
computergestützte Zusammensetzung *f* computerized composition
computergestütztes Bohren *n* **von Leiterplatten (Platinen)** computerized PC board drilling
computergestütztes Erkennen *n* machine-aided cognition
Computergraphik *f* computer graphics
Computerhierarchiesystem *n* hierarchical computer system
Computerintelligenz *f* artificial intelligence *{heuristic programming}*
computerisierte Übersetzung *f* machine translation
Computerlinguistik *f* computer linguistics
Computerlogik *f* computational [machine] logic
Computernetz *n* computer network
Computernetzarchitektur *f* computer network architecture
Computernetzwerk *n* computer network, network of interconnected computers
Computernetzwerk *n* **mit hierarchisch verteilter Intelligenz** hierarchical-distributed computer network
Computernetzwerk *n* **mit verteilter Intelligenz** distributed computer network
Computersprache *f* computer language
Computersystem *n* **mit direkter Wechselwirkung zwischen Mensch und Maschine** interactive computing system
Computersystem *n* **mit verteilter Intelligenz** distributed-intelligence computer system
computerunabhängig computer-independent
computerunterstützte Planung *f* computer-aided planning
computerunterstützte Unterweisung *f* computer-aided instruction
computerunterstützte Übersetzung *f* computer-aided translation
computerunterstützter Unterricht *m* computer-aided instruction
Computerwörterbuch *n* machine dictionary
Connecting-Maschine *f* connecting machine *{a parallel computer}*
Connection *f* connection
Connection-Matrix *f* connection matrix
Connectionsstärke *f* connection strength
Cortex *m* cortex *{the outer portion of a structure}*

D

Dämon *m* demon
Dämpfung *f* attenuation
DANN THEN
 DANN UND NUR DANN IF AND ONLY IF
 DANN WENN THEN IF
darstellen display/to
Darstellung *f* [re]presentation
Darstellungssprache *f* representation language
Darstellungstheorie *f* representation theory
Dateiliste *f* dictionary
Dateiverzeichnis *n* dictionary
Datenanalyse *f* analysis of data
Datenbank *f* data base, data bank
Datenbanksystem *n* data base system
Datenbankverwaltungssystem *n* data base management system
Datenbasis *f* data base
Datenebene *f* rank
Datenfeld *n* array, field
datengesteuert data-driven
 datengesteuerte Inferenz *f* data-directed inference
Datenhierarchie *f* data hierarchy
Datenkette *f* data chain
Datenmanipulation *f* data manipulation
Datenmanipulationssprache *f* data manipulation language
Datensatz *m* record
Datenstruktur *f* data structure
Datenverarbeitung *f* **mit verteilter Intelligenz** distributed data processing
Datenverarbeitungsanlage *f* computer
Daumenregel *f* rule of thumb, hard-and-fast rule
Deduktion *f* deduction *{predicate logic}*
Deduktionssystem *n* deductive system
Deduktionstheorem *n* deduction theorem
deduktiv deductive
 deduktive Fehleranalyse *f* deductive fault analysis, fault tree analysis
 deduktive Folgerung *f* deductive conclusion
 deduktiver Schluß *m* deduction, deductive inference
 deduktives Expertensystem *n* deductive shell
 deduktives System *n* deductive system
definierbar definable
definieren define/to
definiert defined, definite
 definiertes Ereignis *n* definite event
 definiter Automat *m* definite automaton
Definition *f* definition
Definitionsbereich *m* domain, range of definition
Definitionssprache *f* definition language
Degeneration *f* degeneration
Dehnung *f* extension
Dekomposition *f* decomposition
Dekompositionsalgorithmus *m* decomposition algorithm
Dekompositionsmethode *f* decomposition approach
Deltaregel *f* delta rule *{a learning rule}*
denär denary

Dendrit m dendrite *{input channel of the biological neuron}*
Depression *f* depression
deskriptive Logik *f* descriptive logic
deskriptive Sprache *f* descriptive language
Deskriptor m descriptor, keyword
Deskriptor-Definitionssprache *f* descriptor definition language
determiniert determined, determinate
 determinierte Unordnung *f* deterministic chaos (disorder)
 determinierter Automat m determinate automaton
 determinierter Prozeß m deterministic process
 determiniertes Chaos n deterministic chaos (disorder)
 determiniertes System n determinate system
Determiniertheit *f* determinacy
deterministisch deterministic
 deterministische Optimierung *f* deterministic optimization
 deterministische Simulation *f* deterministic simulation
 deterministischer Algorithmus m deterministic algorithm
 deterministischer Automat m deterministic automaton
 deterministisches Optimierungsproblem n deterministic optimizing problem
 deterministisches Signal n deterministic signal
 deterministisches System n deterministic system
dezentral decentralized
dezentralisierte Intelligenz *f* decentralized intelligence *{lowest level of hierarchical systems}*
Diagnose *f* diagnosis
Diagnoseautomat m diagnosis (diagnostic) computer
Diagnosecomputer m diagnosis computer
Diagnosemaschine *f* diagnostic computer
Diagnoseprogramm n diagnostic program
Diagnostik *f* diagnotics
diagnostizierbar diagnosable
Diagonalmatrix *f* diagonal matrix
Diagramm n diagram, graph
Dialekt m dialect
Dialog m dialog[ue]; interactive session
Dialog m zwischen Computern system-to-system communication
Dialog m zwischen Mensch und Maschine man-machine dialogue
Dialogbetrieb m interactive mode
Dialogcomputersystem n interactive computing system
Dialoggerät n dialog[ue] unit
Dialogsprache *f* dialog[ue] language
Dialogsystem n conversational system, dialog[ue] system
Dialogterminal n conversational terminal
Dialogverarbeitung *f* transaction processing
Diätproblem n diet problem, minimum-cost-of-fodder problem
dichotome sequentielle Suche *f* dichotomic sequential search
dichotomer Klassifikator m dichotomic classifier
Dichotomie *f* dichotomy
dichotomische Suche *f* dichotomizing search

Dichte *f* density
Dichtefunktion *f* density function
Differentialgleichung *f* **zweiter Ordnung** second-order differential equation
Differenz *f* difference
Differenzverringerung *f* difference reduction
Digraph *m* directed graph
Dimension *f* dimension *{number of components of a vector}*
direkte Optimierung *f* direct optimization
disjunkt disjoint
  disjunkte Menge *f* disjoint set
Disjunktion *f* contravalence, disjunction, exclusive OR, inclusive OR
disjunktiv disjunctive
  disjunktive Normalform *f* standard sum *{Boolean algebra}*
diskret discrete
  diskrete Optimierung *f* discrete optimization (optimizing)
  diskrete Restlogik *f* remaining discrete logic
  diskreter Automat *m* discrete automaton
  diskreter deterministischer Mehrstufenentscheidungsprozeß *m* discrete deterministic multistage decision process
  diskretes Maximumprinzip *n* discrete maximum principle
Diskretisierung *f* truncation
Diskretisierungsfehler *m* truncation error
Diskretzeitmodell *n* discrete-time model
Distributionsproblem *n* distribution problem, transport[ation] problem

distributives Gesetz *n* distribution (distributive) law
Distributivität *f* distributivity
divergent divergent
divergierend divergent
  divergierende Reihe *f* divergent series
Domäne *f* domain
dominierender Eingang *m* overriding input
Doppelgraph *m* dual graph
doppelt dual
Dreiebenennetzwerk *n* three level network
Dreiebenenperceptron *n* three-layer perceptron
Dreipersonenspiel *n* three-person game
dreistellig ternary *{e.g. relation}*
dreiwertige Logik *f* three-valued (ternary, three-state) logic
Dreizustandslogik *f* *s.* dreiwertige Logik
Dringlichkeit *f* **der Forderung** demand priority *{queueing theory}*
Druckgefäß *m* accumulator
dual dual, binary
  duale Aufgabe *f* **der konvexen Optimierung** dual problem of convex optimization
  duale Boolesche Funktion *f* dual Boolean function
  duale Operation *f* dual operation
  dualer Simplexalgorithmus *m* dual simplex algorithm
  duales Problem *n* dual problem
  duales Zahlensystem *n* binary [number] system
Dualität *f* duality
Dualitätsprinzip *n* duality principle
Dualproblem *n* dual problem

Duell n duel *{theory of games}*
Duellsituation *f* duel condition *{theory of games}*
dummes Terminal *n* dumb terminal
Durchfluß *m* flow
durchführbar performable
Durchführbarkeit *f* feasibility, performability
durchführen perform/to
Durchschnitt *m* intersection *{set theory}*
durchschnittlich average
durchschnittsfremd disjoint *{set theory}*
Durchschnittsmenge *f* intersection set
dyadische Boolesche Operation *f* dyadic Boolean operation
dynamisch dynamic
dynamische Stabilität *f* dynamic stability *{ability of a network to remain within its functional boundaries}*
dynamische Wissensbank *f* dynamic knowledge base, short-term memory, working memory
dynamisches Modell *n* dynamic model
dynamisches Prozeßmodell *n* dynamic process model
dynamisches Spiel *n* dynamic game

E

E-Fläche *f* E surface *{three-dimensional curved surface representing the energy states within a neural network}*
E-Funktion *f* E function
E-Netz *n* evaluation net *{Petri net}*

Ebene *f* plane
ebener Graph planar (plane) graph
echt real, proper
echte Teilmenge *f* proper subset
echtes Problem *n* real-world problem
Editor *m* editor *{software tool}*
Effekt *m* effect
effektiv effective, actual, real
Effektivität *f* effectiveness
Effektivität *f* **des Innovationsprozesses** efficiency of an innovation process
Effektivitätskriterium *n* criterion of effectivity
Effektor *m* [end] effector, actuating unit
Effizienz *f* efficiency
Eigenentropie *f* inherent entropy
Eigenfunktion *f* proper function
Eigeninstabilität *f* intrinsic instability
Eigenschaft *f* property; feature
Eigenschaftsdetektor *m* feature detector *{a group of neurons who recognize a particular feature}*
Eigenschaftshierarchie *f* inheritance hierarchy
Eigenschaftsliste *f* property list, frame
Eigenstabilität *f* inherent stability
Eigenstabilitätsgrenze *f* natural-stability limit
eigentliche Untergruppe *f* proper subgroup
Eigenwert *m* characteristic value
einbetten embed/to
einblättrige Fläche *f* single-leaf surface
einblättrige Phasenebene *f* single-leaf phase plane

einblättrige Zustandsebene *f* single-leaf state plane
eindeutig unique *{mathematics}*
eindeutige Funktion *f* injective function
eindeutiger Zusammenhang *m* non-ambiguous relation
Eindeutigkeit *f* uniqueness
Eindeutigkeitsoperator *m* uniqueness operator
Eindeutigkeitssatz *m* uniqueness theorem
eindimensionale Sprache *f* one-dimensional language
Einebenenstruktur *f* single-level structure
eineindeutige Beziehung *f* one-to-one relation
eineindeutige Funktion *f* one-to-one function
einengen narrow/to
einfach simple
einfach zusammenhängend simply connected *{point set}*
einfache Funktion *f* simple function
einfache Gruppe *f* simple group *{group theory}*
einfaches Spiel *n* simple game *{game theory}*
einfache Vererbung *f* simple inheritance
Einfluß *m* influence
Einfluß *m* **der Zuverlässigkeit bei Expertensystemen** impact of reliability in expert systems
Eingabealphabet *n* input alphabet
Eingabeauflösung *f* input resolution *{numerical control}*
Eingabeelement *n* input primitive *{logic}*

Eingabeneuron *n* input neuron *{receives data from the outside world}*
Eingabewert *m* input [value]
Eingang *m* arrival
Eingangsgröße *f* input [quantity, variable]
Eingangslogik *f* input logic
Eingangssignal *n* input [signal]
Eingangsstrom *m* arrival stream *{queueing theory}*
Eingangsvalenz *f* input valency *{graph}*
Eingangszeit *f* arrival time
eingebaut internal
eingebettete Markow-Kette *f* embedded Markov chain
eingebettetes Expertensystem *n* embedded expert system *{expert system which receives its input from sensors and delivers its output to actuators}*
eingeschränkt constrained
eingeschwungen stationary
eingestellt adjusted
Eingriff *m* **durch den Menschen** human intervention
einhalten control/to *{set value}*
Einheit *f* entity *{member of a class of objects with the same properties, e g. a human being}*
Einkantengraph *m* single-edge graph
einlesen teach in/to *{movement of robot arm}*
Einpunkt-Randwertaufgabe *f* single-point boundary value problem
Einrichtbetrieb *m* setting-up mode *{numerically controlled machine tool}*
Einrichtungen *fpl* hardware

Einsatzbereich *m* range
einschalten activate/to; actuate/to
Einschalten *n* actuation
Einschaltung *f* activation
einschließende ODER-Funktion *f* inclusive OR function
einschließendes ODER *n* disjunction, inclusive OR
einschränken control/to {interference}
Einschränkung *f* restriction
Einschrittalgorithmus *m* single-step algorithm
Einselement *n* monoid
Einsquantor *m* existential quantifier
einstellbarer Schwellenwert *m* adjustable threshold [value]
Einstellelement *n* adjustment [element]
einstellen control/to, adjust/to; trim/to
einstelliges Prädikat *n* unary predicate
Einstellung *f* adjustment
Einstichverfahren *n* dichotomizing search
einstufig single-stage
einstufige Entscheidung *f* single-stage decision
einstufiger Entscheidungsprozeß *m* single-stage decision process
eintasten keyboard/to
eintippen keyboard/to
Eintrag *m* field
eintreffende Forderung *f* arriving demand {queueing theory}
eintreten happen/to {event}
Eintrittswahrscheinlichkeit *f* occurrence probability
Einüben *n* rehearsal {memorization through practice or repetition}

Einwirkung *f* action; effect
Einzelfall *m* instantiation
Einzelteil *n* part
Einzielfunktion *f* single-objective function
Einzielstruktur *f* single-objective structure
Einzielsystem *n* single-objective system
Einzweckroboter *m* special[-type] robot
Element *n* [logic] element; literal
elementar elementary
Elementaralternative *f* elementary alternative, maxterm
Elementarautomat *m* elementary automaton
elementare Assoziation *f* [elementary] association
elementare Beziehung *f* elementary assertion (fact, relationship)
elementare Kette *f* elementary chain {graph}
elementare logische Verknüpfung *f* elementary logical connection
elementare Sprache *f* elementary language
elementare Wortoperation *f* arithmetical logical operations on computer words
elementarer Automat *m* elementary automaton
elementarer Zyklus *m* elementary cycle {graph theory}
elementares System *n* elementary system
Elementarentscheidung *f* elementary decision
Elementarereignis *n* elementary (simple) event

Elementarglied *n* basic element
Elementarkonjunktion *f* elementary conjunction, minterm
Elementaroptimierung *f* elementary optimization
Elementarsystem *n* elementary system
Elementarterm *m* elementary term
Elementarvernüpfung *f* elementary connection
Elementarzeichen *n* elementary character
Elementarzyklus *m* elementary cycle {graph theory}
elementfremd disjoint
Eliminierbarkeit *f* eliminability
eliminieren eliminate/to
eliminierende Suche *f* dichotomizing search
Empfänger *m* information receiver
Empfangsneuron *n* receiving neuron
Empfangsort *m* destination
empfindlich gegen zufällige Ereignisse accident-sensitive
Empfindlichkeitskurve *f* sensitivity curve
Empfindlichkeitsschwelle *f* sensitivity threshold
empirische Formel *f* empiric[al] formula
empirische Lösung *f* empiric[al] solution
empirisches Modell *n* empiric[al] model
Ende *n* end, termination
Endereignis *n* final event {network}
Endkante *f* end edge {tree}
endlich finite
   endlich-dimensional finite dimensional
   endliche Gruppe *f* finite group
   endlicher Akzeptor *m* finite state acceptor
   endlicher Automat *m* finite automaton, finite (sequential) state machine
   endlicher Graph *m* finite graph
   endliches Spiel *n* finite game
Endpunkt *m* end point {vector}
Endstellung *f* end point
Endwert *m* final value, value of end point
Endzustand *m* final state, finite condition
Energiefunktion *f* energy function {defines the connection matrix and the initial input and describes the energy state of the network}
Energieminimum *n* energy minimum {stable state of the energy surface}
Engset-Verkehr *m* source traffic
entarten degenerate/to
entartetes Spiel *n* degenerated game
entartetes System *n* degenerated system
Entartung *f* degeneration
entfernen dequeue/to {from queue}
entgegengerichtet inverse
enthalten hold/to
Enthaltenseinsrelation *f* inclusion relation
entkoppeltes System *n* non-interacting (decoupled) system
Entkopplung *f* decoupling, non-interaction
Entkopplungskondensator *m* neutralizing capacitor
entlocken clear/to
entregende Synapse *f* inhibitory synapse {causes a decrease in the activation level of the receiving neuron}

Entropie f average information content, [information] entropy
entscheidbar decidable
entscheidbarer Ausdruck m decidable expression
Entscheidbarkeit f decidability
Entscheidbarkeitsproblem n decidability problem
entscheiden decide/to
entscheidendes System n decision system
Entscheidung f decision, arbitration
Entscheidung f bei Gewißheit decision with certainty
Entscheidung f bei Risiko decision with risk
Entscheidung f bei Ungewißheit decision with uncertainty
Entscheidungsablauf m decision sequence
Entscheidungsalgorithmus m decision algorithm
Entscheidungsanalyse f decision analysis
Entscheidungsbaum m decision tree
Entscheidungsbaumverfahren n decision tree method
Entscheidungsbefehl m decision instruction
Entscheidungsbereich m decision domain
Entscheidungsebene f decision level
Entscheidungseinheit f decision device
Entscheidungselement n decision element (gate)
Entscheidungsfindung f decision finding, decision making
Entscheidungsfindung f unter den Bedingungen der Unbestimmtheit decision assumption at conditions of indetermination
Entscheidungsfolge f decision sequence
Entscheidungsfunktion f decision function
Entscheidungsgatter n decision element
Entscheidungsgehalt m decision content
Entscheidungsglied n decision element
Entscheidungsgraph m decision graph
Entscheidungsgröße f decision variable
Entscheidungshilfe f decision support system
Entscheidungsinhalt m decision content
Entscheidungsinstruktion f decision instruction
Entscheidungsintegrator m decision integrator, saturating integrator
Entscheidungskriterium n criterion of judging, decision criterion
Entscheidungslogik f arbitration logic, deciding logic
Entscheidungsmatrix f decision matrix, connectivity matrix, connection
Entscheidungsmethode f decision method
Entscheidungsmodell n decision model
Entscheidungsnetz n decision network
Entscheidungsnetzplan m decision network
Entscheidungsprinzip n decision principle
Entscheidungsproblem n decision problem

Entscheidungsprogramm *n* decision program
Entscheidungsprozedur *f* decision procedure
Entscheidungsprozeß *m* decision process
Entscheidungsraum *m* decision space
Entscheidungsregel *f* decision rule
Entscheidungsschaltung *f* decision circuit; decision element
Entscheidungsschwelle *f* decision threshold
Entscheidungsstelle *f* decision location *{Petri net}*
Entscheidungsstrategie *f* decision strategy
Entscheidungsstufe *f* decision stage
Entscheidungssystem *n* decision system
Entscheidungstabelle *f* decision table
Entscheidungstheorie *f* decision theory
entscheidungsunterstützendes System *n* decision support[ing] system
Entscheidungsvariable *f* decision variable
Entscheidungsvektor *m* decision vector
Entscheidungsverfahren *n* decision procedure
Entscheidungswahrscheinlichkeit *f* decision probability
Entstehungsgeschwindigkeit *f* **von Nachrichten** average source entropy
ENTWEDER-ODER *n* exclusive OR, OR-ELSE
ENTWEDER-ODER-Gatter *n* exclusive OR circuit (element, gate, network)
ENTWEDER-ODER-Schaltung *f* exclusive OR circuit (element, gate, network)
ENTWEDER-ODER-Tor *n* ENTWEDER-ODER-Tor n
entwerfen design/to
Entwicklungsmodell *n* developmental model
Entwicklungswerkzeug *n* development tool
Entwurf *m* design
Entwurfsalgorithmus *m* design algorithm
Entwurfsautomatisierung *f* design automation
Entwurfsstrategie *f* design strategy
Entwurfstheorie *f* design theory
erben inherit/to
erbliche Eigenschaftshierarchie *f* inheritance hierarchy
Erbsystem *n* hereditary system
ereignen/sich happen/to
Ereignis *n* event, occurrence
Ereignisalgebra *f* algebra of events, event algebra
Ereignisfeld *n* event field, field of events
Ereignisfolge *f* event sequence
ereignisgesteuert data-driven, event-driven, bottom-up-driven
ereignisgesteuerte Logik *f* event-driven logic
ereignisgesteuerte Steuerstruktur *f* data-driven control structure, event-driven control structure, bottom-up control structure
Ereigniskompatibilität *f* event compatibility
Ereignismarke *f* event mark

ereignisorientierte Simulation *f*
  [critical-] event simulation
ereignisorientierter Netzplan *m*
  event-oriented network
Ereignisraum *m* event space
Ereignisspeicher *m* memory of
  events
Ereigniswahrscheinlichkeit *f*
  occurrence probability
Erfahrung *f* experience, expertise;
  brainware *{programming}*
Erfahrungsspeicher *m* experience
  store
Erfahrungswert *m* empiric[al] value
Erfahrungswissen *n* experiential
  knowledge, heuristic knowledge,
  surface knowledge
erfassen 1. cover/to; 2. perceive/to
  {pattern}; keyboard/to *{text}*
Erfassen *n* perception *{pattern}*
Erfolg *m* success
Erfolgsbilanzierung *f* success balancing
Erfolgsfunktion *f* success function
Erfolgskriterium *n* success criterion
Erfolgsspinne *f* polar coordinate
  success pattern *{decision theory}*
erfüllbar performable
Erfüllbarkeit *f* satisfiability,
  performability *{logic expression}*
erfüllen satisfy/to; perform/to
Erfüllung *f* von Randbedingungen
  constraint satisfaction
Ergodentheorie *f* ergodic theory
ergodischer Prozeß *m* ergodic
  process
ergodischer Zustand *m* ergodic
  state
ergodisches System *n* ergodic system
erhalten obtain/to

erhältlich available
Erhaltung *f* conservation *{mathematics}*
Erhaltung *f* des Folgerungswegs
  [in Wissensbasen bei Hinzukommen neuer Information] reason
  maintenance
Erhaltungssatz *m* conservation law
Erinnerungsvermögen *n* memorization
erkennbar perceptible, recognizable
Erkennbarkeit *f* detectability, intelligibility, perceptibility, recognizability
erkennen recognize/to
Erkennen *n* cognition
Erkennen-Handeln-Zyklus *m* recognize-act cycle
erkennende Grammatik *f* recognizing grammar
Erkenntnis *f* cognition
Erkenntnismethode *f* cognitive
  approach *{learning approach that
  focusses on the modification of cognitions by experience}*
erkenntnisphilosophisches
  Modell *n* cognitive-philospical
  model
Erkenntnisprozeß *m* cognitive
  process
Erkenntniswissenschaft *f* cognitive
  science
Erkennung *f* [pattern] recognition
  {s.a. Spracherkennung}
Erkennung *f* visueller Objekte
  recognition of visual objects
Erkennungsalgorithmus *m* identifying algorithm
Erkennungsgerät *n* recognizer
Erkennungsspeicher *m* recognition
  memory

Erkennungssystem *n* identification system, identifying system; recognition system *{pattern recognition}*
Erkennungszuverlässigkeit *f* reliability of recognition
erklärend declarative
erklärende Wissensdarstellung *f* declarative knowledge representation
Erklärung *f* explanation; statement
Erklärung *f* **mitten im Ablauf** midrun explanation *{during program run}*
Erlang-Verteilung *f* Erlang distribution
Erläuterung *f* explanation
Ermittlung *f* determination
Ermittlungsmodell *n* ascertainment model
Ernennungsproblem *n* assignment problem, coordination problem
Erneuerungstheorem *n* renewal theorem
Erneuerungstheorie *f* recovery (renewal, replacement) theory
erproben try/to
Erprobung *f* trial
errechnen evaluate/to
erregende Synapse *f* excitatory synapse *{causes an increase in the activation level of the receiving neuron}*
Erregung *f* excitation, input
erreichbar available
Ersatz *m* substitution
Ersatztheorie *f* renewal (replacement) theory
erschöpfende Suche *f* exhaustive search *{all paths of the tree are searched for}*
Ersetzbarkeitstheorem *n* replaceability theorem

ersetzen durch substitute by/to
Ersetzungsaxiom *n* substitution axiom
Erwartung *f* expectation, expectancy
erwartungsgeführter Lösungsweg *m* expectation-guided solution approach
erwartungsgesteuert expectation-driven
erwartungsgesteuerte Schlußfolgerung *f* expectation-driven reasoning
Erwartungswert *m* mathematical expectation [value]
erweiterbare Struktur *f* expandable structure
erweiterte Simplexmethode *f* extended simplex method
erweitertes Modell *n* expanded model
erweiterungsfähig expandable
Erweiterungskörper *m* extension field
erzeugen create/to; generate/to *{e.g. curve}*
Erzeugung *f* generation *{e.g. program}*
Erzeugung *f* **von Regellosigkeit** randomization
ES GIBT EIN THERE IS *{predicate logic}*
Existentialaussage *f* existential proposition
existentieller Quantifikator *m* existential quantifier
Existenz *f* existence
Existenzaussage *f* existence sentence
Existenzfunktor *m* existential quantifier

Existenzoperator *m* existential quantifier
Existenzquantifikator *m* existential quantifier
Existenzsatz *m* existence theorem
Existenztheorem *n* existence theorem
Exklusion *f* exclusion
exklusiv exclusive
Exklusiv-ODER *n* exclusive OR, EXOR [operation], contravalence, OR-ELSE *{s.a.* ausschließendes ODER*}*
Experte *m* expert
Experte *m* **auf Spezialgebiet** domain expert
Expertenschätzung *f* expert estimation
Expertensystem *n* expert [system], knowledge information processing system
Expertensystem *n* **mit Beispielen als Wissensbasis** induction system, example-driven system
Expertensystemschale *f* expert shell
Expertensystemshell *n* expert shell
Expertenwerkzeug *n* expert tool
Expertenwissen *n* expertise; expert knowledge
Extension *f* extension
extensionale Logik *f* extensional logic
Extensionalitätsaxiom *n* axiom of extensionality
Extensionalitätsprinzip *n* extensionality principle
externes Objekt *n* external entity *{source or sink}*
Extremale *f* extremal *{curve}*
Extremalproblem *n* extremal problem

Extremalsystem *n* extremal system, extremum-searching adaptive system
Extremalwert *m* extremal value, extremum [value]
extremieren extremize/to *{minimizing or maximizing}*
Extremierungskriterium *n* extremizing criterion
Extremum *n* extremal value, extremum [value]
Extremwert *m* extreme value

F

Facette *f* facet *{field for entries into a frame}*
Fach *n* slot *{frame element}*
Fachwissen *n* expertise
Fähigkeit *f* capability, ability
Fähigkeit *f* **des Gehirnes** brain capability
faires Spiel *m* fair game
Fakt *m* fact
Faktendatenbank *f* working memory
Fall *m* case
Fallbetrachtung *f* instantation
Fallstudie *f* case [study]
Falltürfunktion *f* trapdoor function
Fallunterscheidung *f* definition by cases
falsch false, untrue *{logic}*
falsche Aussage *f* false declarative sentence
falsche Folgerung false conclusion
Fälschung *f* forgery
Faltung *f* convolution
Faltung *f* **einer Funktion mit sich selbst** convolution of a function with respect to itself

Faltung *f* **von Verteilungen von Zufallsvariablen**   convolution of distributions of random variables
Faltungsalgorithmus *m*   convolution algorithm
Faltungsintegral *n*   convolution integral
Faltungsprodukt *n*   convolution product
Faltungssatz *m*   convolution theorem
Faltungssumme *f*   convolution sum
Farbspritzroboter *m*   spray-painting robot, paint-spraying robot
FAST IMMER   NEARLY ALWAYS   *{fuzzy quantor}*
FAST NIE   NEARLY NEVER   *{fuzzy quantor}*
Faustregel *f*   hard-and-fast rule, heuristic rule, rule of thumb
Fehlentscheidung *f*   false decision
Fehler *m*   error; mistake
  Fehler *m* **bei der Diskretisierung analoger Werte**   truncation error
  Fehler *m* **des Entscheidungselementes**   error of the decision element
  Fehler *m* **des Modells**   model error
  Fehler *m* **durch Approximation**   approximation error
Fehleranalyse *f* **durch Induktion (induktiven Schluß)**   inductive fault analysis
Fehlerauffindung *f*   error detection
Fehlerbaum *m*   fault tree
Fehlerbaumanalyse *f*   deductive fault analysis, fault tree analysis
fehlerbehaftet   noisy
Fehlerbeseitigung *f*   error debugging
Fehlerbeseitigung *f* **in der Primärsprache**   source-language debugging

Fehlerdiagnose *f*   error diagnostics, fault diagnosis
Fehlerkorrektur *f*   debugging *{e.g. in a program}*
Fehlerkorrektursystem *n*   error-correcting system
Fehlerkorrigierbarkeit *f*   error correctability
Fehlersuchbaum *m*   trouble-shooting tree
Fehlersuche *f*   trouble tracking
Fehlertoleranz *f*   fault tolerance   *{ability of a network to keep processing, possibly with degradation, when a small number of neurons are disabled}*
Fehlerzurückverfolgung *f*   error backpropagation
Fehlmengenkosten *pl*   shortage cost
Fehlverhalten *n*   faulty behaviour
Feld *n*   field, array
Feld *n* **der Verhaltenslinien**   field of lines of behaviour
Feldausdruck *m*   field expression
Feldgröße *f*   field quantity
feldprogrammierbare logische Folgesteuereinheit *f*   field-programmable logic-sequence
Feldprozessor *m*   array processor
Feldvereinbarung *f*   array declaration
Fensterfunktion   window function
Fernbedienungseinrichtung *f*   teleoperator *{part of robot system}*
Fertigung *f*   production
Fertigungsstraße *f*   production (process) line
festeingestelltes Bezugsmodell *n*   fixed reference model
Festigen *n*   rehearsal *{memorization through practice or repetition}*

festprogrammierter Roboter *m* fixed-sequence robot
Feststellung *f* determination
festverdrahtete Logik *f* hard-wired logic
Feuerhäufigkeit *f* firing frequency *{measure of the activity of a neuron, measured by the number of impulses per second}*
feuern fire/to *{rule}*
Feuerneigung *f* excitatory tendency *{tendency of a synapse or connection to cause firing of the receiving neuron}*
Fibonacci-Suche *f* Fibonacci search
Fibonacci-Suchprozeß *m* Fibonacci search process
Fibonacci-Zahl *f* Fibonacci number
Figur *f* pattern
fiktive Aktivität *f* fictitious activity
fiktiver Spieler *m* fictive player
fiktiver Vorgang *m* fictitious activity
fiktives Spiel *n* fictive game
Filter *n* **mit begrenztem Gedächtnis** finite-memory filter
Filter *n* **mit lokaler Kopplung** local-coupled filter
Finger *m* finger *{part of robot}*
finit finite
Finite-Elemente-Modell *n* finite-element model
finiter Graph *m* finite graph
Fläche *f* area
Fließbandverarbeitung *f* flow line processing, pipelining
Flugbahn *f* trajectory
Fluß *m* flow
Flußgraph *m* flow graph
Folge *f* sequence *{mathematics}*
folgegesteuert sequence-controlled

folgern conclude/to
Folgerung *f* conclusion, consequent, inference
Folgerungscomputer *m* inference machine
folgerungserblich conclusion-hereditary, hereditary under conclusion
Folgerungsmenge *f* set of conclusions
Folgerungsnetz *n* inference net
Folgerungsoperation *f* inference [operation]
Folgerungsrelation *f* conclusion relation
Folgerungsschritt *m* inference step
Folgerungsstrategie *f* inference strategy
Folgeschaltsystem *n* sequential state machine
Folgestruktur *f* follow-the-leader structure
Folgezustand *m* consequence, follow-up state
Follow-the-leader-Struktur *f* follow-the-leader structure
Foodscher Algorithmus *m* Food's algorithm
Foodscher Funktor *m* Food's algorithm
Forderung *f* statement; demand, requirement
Forderungspriorität *f* demand priority
Forderungsstrom *m* demand stream, stream of demands (requirements) *{queueing theory}*
Form *f* pattern; mode; gestalt *{pattern perceived as a whole}*
formale Aussage *f* formal statement
formale Logik *f* formal (symbolic) logic

formale Operation *f* formal operation
formale Sprache *f* formal language
formaler Parameter *m* formal parameter
formales Modell *n* formal (symbolic) model
formalisieren formalize/to
formalisierte Sprache *f* formalized language
Formalismus *m* formalism
Formblatt *n* form
Formel *f* formula
Formelübersetzer *m* formula translator
Formular *n* form
Fortbewegung *f* **auf Beinen** legged locomotion *{robot}*
fortgeschritten advanced
fortlaufend consecutive
Fortschritt *m* progress
Fortsetzung *f* continuation
Frage *f* question
fragen inquire/to
Fragenbeantwortung *f* question answering
Fragenbeantwortungssystem *n* question answering system
Fraktal *n* fractal
frei free
  freie Interpretation *f* free interpretation
  freie Optimierung *f* constraint-free optimization
  freie Variable *f* free variable
  freier Automat *m* free automaton
  freier Endzustand *m* unconstrained finite state
freigeben clear/to
Freitextretrieval *n* free text retrieval

Fuge *f* joint *{between adjacent parts}*
Fügeroboter *m* joining robot
Fundamentalalternative *f* fundamental alternative
Fundamentalkonjunktion *f* fundamental conjunction
Fundamentalterm *m* fundamental term
Fundierungsaxiom *n* axiom of regularity *{set theory}*
Fundstelle *f* hit
Fünffarbensatz *m* five-colour theorem *{theory of graphs}*
Funktion *f* function *{mathematics}*
funktionelle Anwendung *f* functional application *{problem solving}*
funktionieren run/to
Funktionsanalogie *f* analogy of functions
funktionsbereit active
Funktionsdiagramm *n* action chart
Funktionssimulator *m* function simulator
Funktionstabelle *f* **für Wahrheitswerte** truth table
Funktionstafel *f* **für Wahrheitswerte** truth table
Funktionsweise *f* function
Funktor *m* functor
fusionieren merge/to

G

gabeln/sich bifurcate/to
Gabelung *f* bifurcation
Galton-Brett *n* Galton board
Ganglion *n* ganglion *{collections of nerve cells being just outside the autonomous nervous system}*

ganzzahlige 158

ganzzahlige Optimierung *f* discrete optimization (optimizing)
Gatter *n* gate
Gatterschaltung *f* gate circuit
geballtes Wissen *n* compiled knowledge
Gebot *n* bid {*of a rule*}
gebundener Endzustand *m* constrained finite state
Geburts-und Todesprozeß *m* birth-and-death process
Geburtsprozeß *m* birth process
Gedächtnis *n* memory
gedächtnislos memory-free, memoryless
Gedächtnissystem *n* hereditary system
Gedanke *m* thought
Gedankenverarbeitung *f* thought processing
Gedankenversuch *m* imaginary exercise, thought experiment
gedruckte Information *f* paper-based information
Geduldzeit *f* patience time {*queueing theory*}
gefundene Stelle *f* hit
Gegenbeispiel *n* counterexample, negative example
Gegenimplikation *f* replication
gegenseitige Abhängigkeit *f* interdependence
gegenseitige Beeinflussung *f* interaction
gegenseitiger Ausschluß *m* mutual exclusion
Gegenströmung *f* counterflow
Gegenwart *f* presence
Gehalt *m* content
Geheimhaltung *f* **von Information** information hiding

Geheimverschlüsselung *f* encryption
Gehirn *n* brain
Gehirnfunktion *f* brain function
gekoppelt linked
Gelenkarmroboter *m* jointed-arm robot
Gelenk *n* joint, linkage {*robot*}
Gelenkgeometrie *f* articulated geometry {*robotics*}
gelöster Konflikt *m* solved conflict
gelten hold/to {*rule*}
gemeinsame Entropie *f* [an Kanaleingang und Kanalausgang] joint entropy
gemeinsame Wahrscheinlichkeit *f* joint probability
gemischt mixed
gemischte Strategie *f* mixed strategy
gemischter Graph *m* mixed graph
gemischtes System *n* mixed system
gemittelt averaged
GENAU DANN [WENN] IF AND ONLY IF
GENAU DANN WENN THEN AND ONLY THEN
Genauigkeit *f* **der Auflösung** accuracy of resolution
generalisieren generalize/to
generalisierte Koordinaten *fpl* generalized coordinates
Generalisierung *f* generalization [rule]
Generation *f* generation {*technical development*}
generative Grammatik *f* generating grammar
Generator *m* generator
generieren generate/to, create/to

generische Eigenschaft *f* generic property
generische Grammatik *f* generative grammar
generische Routine *f* generic routine
generische Situation *f* stereotyped situation
genetischer Algorithmus *m* genetic algorithm
generisches Programm *n* generic routine
geometrische Information *f* geometrical information
geordnet ordered
  geordnete Automatisierung *f* sequenced automation
  geordnete Menge *f* ordered set
  geordneter Baum *m* ordered tree
  geordnetes Paar *n* ordered pair *{set theory}*
geradlinig rectilinear
  geradlinige Koordinaten *fpl* rectilinear coordinates
Geräte *npl* hardware
Geräteteil *n (m)* subset
gerichtet directed, oriented
  gerichtete Kante *f* directed edge
  gerichtete Pfeilfolge *m* oriented walk
  gerichteter Graph *m* directed graph
  gerichteter Graph *m* directed graph, oriented graph
  gerichteter Weg *m* oriented path, oriented trail
geringhalten control/to *{interference}*
GERT-Methode *f* graphic evaluation and review technique *{network}*
Gerüst *n* framework *{graph}*

Gesamtausfall *m* total malfunction
Gesamtheit *f* totality
Gesamtstabilität *f* overall stability
Gesamtstrategie *f* overall strategy
Gesamtsystem *n* overall system, total system
Geschichte *f* **der künstlichen Intelligenz** history of AI
geschichtete Sprache *f* stratified language
geschlossene Kette *f* cycle
geschlossener Kreis *m* circuit
geschlossener Zyklus *m* closed cycle
geschriebene Information *f* paper-based information
Gesetz *n* law
  Gesetz *n* **von der Existenz des Nullelements** null-class law
Gesichtspunkt *m* objective
gespeicherte Daten *npl* record
gespeicherte Logik *f* stored logic
gespiegeltes Metaspiel *n* mirrored metagame
Gespräch *n* session
Gestalt *f* gestalt *{pattern perceived as a whole}*
Gestalterkennung *f* pattern recognition
gestört disturbed
  gestörte Information *f* disturbed information
  gestörter Zustand *m* disturbed state
gesunder Menschenverstand *m* common sense
Gewicht *n* weight *{value assigned on each connection at the input of a neuron; controls how much of the incoming signal is passed on to the neuron}*

gewichtete Matrix *f* weighted matrix
Gewichtsfaktor *m* weight
Gewichtsmatrix *f* connection (connectivity, weight) matrix *{matrix of the connection weights in a Hopfield network}*
Gewinn *m* gain, profit
gewinnen/einen Extremwert obtain an extremum/to
Gewinnfunktion *f* gain function, profit function
Gewinnmatrix *f* gain matrix, profit matrix
Gewinnmaximierung *f* gain maximization
Gewinnoptimierung *f* gain optimization
Gewißheit *f* [confirmative] certainty
Gewöhnung *f* habituation *{diminished responsiveness to a stimulus when it becomes expected}*
gieren yaw/to *{movement of a robot}*
gleichförmig steady
Gleichgewicht *n* balance
gleichgewichtiger Code *m* symmetric[al] balanced code
Gleichgewichtsbedingung *f* balance condition, steady-state condition
Gleichgewichtsoptimierung *f* steady-state optimization
Gleichgewichtssituation *f* balance situation
Gleichgewichtsstrategie *f* balance strategy
Gleichgewichtszustand *m* balanced state, static state
Gleichheit *f* equality
gleichmächtig equipotent *{set}*
  gleichmächtige Menge *f* equipotent set
gleichmäßig steady

gleichmäßige Approximation *f* **einer Funktion** regular approximation of a function
gleichsetzen equate/to
Gleichung *f* equation
  Gleichung *f* **in Boolescher Algebra** Boolean equation
Gleichungslöser *m* equation solver
Gleichverteilung *f* equipartition
Gleichverteilungseigenschaft *f* equipartition property
gleichwahrscheinlich equiprobable
gleichwertig equivalent
gleichzeitig concurrent, simultaneous
  gleichzeitige Arbeitweise *f* concurrent mode
  gleichzeitige Betätigung *f* simultaneous operation
  gleichzeitige Verarbeitung *f* parallel processing
  gleichzeitiges Arbeiten *n* simultaneous operation
Gleichzeitigkeit *f* simultaneity
Gleichzeitigkeitslogik *f* concurrency logic
gleitender Zustand *m* floating state
Glied logic element; link *{network}*; element *{part of system}*; term *{function}*
global global
  globale Datenbasis *f* global data base
  globale Lösung *f* global solution
  globale Stabilität *f* global stability
  globale Suche *f* global search
  globale Variable *f* global variable
  globales Maximum *n* global maximum
  globales Minimum *n* global minimum

globales Optimum *n* global optimum
globales Extremum *n* global extreme value
Glücksspiel *n* gamble, game of hazard, hazard game
Gödelsches Unvollständigkeitstheorem *n* Goedel's theorem of incompleteness
Grad *m (n)* degree (e.g. circle or temperature]
grad grad {gradient}
Gradient *m* gradient {maximum rate of change in a function}
Gradientenmethode *f* gradient (ascent) method, hill-climbing method
Gradientenprojektionsmethode *f* gradient projection method
Gradientensuche *f* gradient search
Gradientenvektor *m* gradient vector
Gradientenverfahren *n* gradient technique
Grammatik *f* grammar
Graph *m* graph
 Graph *m* **mit Schleifen** looped graph
 Graph *m* **ohne Schleifen** loopless graph
Graphenstruktur *f* graph structure
Graphenstrukturrepräsentation *f* graph structure representation
Graphenstrukturtransformation *f* graph structure transformation
Graphentheorie *f* graph theory
Graphik *f* graphics
Graphiksprache *f* graphic[s] language
 graphische Anzeige *f* graphics
 graphische Darstellung *f* graphic representation
 graphische Manipulationssprache *f* graphics manipulating language
Greifer *m* **mit mehreren Fingern** multifingered hand {robot}
Grenzbedingung *f* boundary condition
Grenze *f* **der natürlichen dynamischen Stabilität** natural transient stability limit
Grenzpunkt *m* accumulation point
Grenzwertproblem *n* boundary value problem
Grenzwertregelung *f* adaptive control with constraints
Größe *f* amount, parameter
Größenordnung *f* order
Grundbaustein *m* basic element
Grundelement *n* basic element
Grundfrequenz *f* fundamental frequency
Grundfunktion *f* basic function
Grundgatter *n* basic gate
Grundgesetz *n* fundamental rule
Grundglied *n* basic element
Grundmenge *f* basic set {mathematics]
Grundmodell *n* **der Entscheidung** basic concept of decision
Grundproblem *n* primal problem {simplex method}
Grundrißoptimierung *f* layout optimization
grundsätzlicher Lösungsweg *m* fundamental approach
Grundstruktur *f* basic structure
Grundsymbol *n* fundamental symbol
Grundverknüpfung *f* basic interconnection (connective) {logic}
Gruppentheorie *f* group theory
gültig valid, legitimate

Gültigkeit 162

gültige Aussagenverbindung *f* valid argument
gültiger Ausdruck *m* valid expression
Gültigkeit *f* validity
Gültigkeitsbereich *m* domain [of validity]
Gültigkeitserklärung *f* validation
Gültigkeitskontrolle *f* validity check[ing] (test)
Gültigkeitsprüfung *f* validity check[ing] (test)
günstigster Fall *m* best case
Güte *f* der Anpassung goodness of fit
Güteklasse *f* class
Gütekriterium *n* optimality criterion

H

Habituation *f* habituation *{diminished responsiveness to a stimulus when it becomes expected}*
halbautomatisch semiautomatic
halbempirische Formel *f* semi-empirical formula
halbempirisches Modell *n* semi-empirical model
halbgeordnete Menge *f* half-ordered set
Halbgruppe *f* semi-group
Halbgruppentheorie *f* half-group theory
Hamilton-Linie *f* Hamilton line *{graph theory}*
Hamming-Abstand *m* Hamming distance *{number of neurons that have different states in two different sets of memory patterns}*
Hand *f* mit mehreren Fingern multifingered hand
handeln act/to
Handelsreisendenproblem *n* travelling-salesman problem
handhaben manipulate/to, handle/to
Handhabung *f* manipulation
Handhabungsautomat *m* handling machine (robot), industrial robot, manipulator-type robot
Handhabungseinheit *f* pick-and-place unit
Handhabungsorgan *n* end effector
Handhabungsroboter *m* handling machine (robot), manipulation robot, pick-and-place robot
Handhabungssystem *n* handling system
Handlung *f* action
Handlungseingang *m* performing input *{decision table}*
Handlungsfeld *n* action field
Handlungskoalition *f* performing (action) coalition
Hardware *f* hardware
Hardwareebene *f* hardware level
hart hard
Hasard *m* hazard
Hasarderkennung *f* hazard detection
Hasarderscheinung *f* hazard phenomenon
Hasardproblem *n* hazard problem
Hasardspiel *n* game of hazard
Häufung *f* accumulated error
Häufungspunkt *m* accumulation point
Hauptattribut *n* prime attribute
Hauptebene *f* primary plane
Hauptleitung *f* trunk
hauptsächlich primary
Haushaltroboter *m* household robot

Heap *m*   heap *{tree}*
Hebb-Regel *f*   Hebb's rule *{learning rule that states that the synaptic weight changes are proportional to the product so the synaptic activities of both the sending and receiving cells}*
heilen   heal/to
hemmen   inhibite/to
hemmend   inhibitory *{to prevent firing of th receiving neuron}*
Hemmung *f*   inhibition
heraussuchen   retrieve/to *{stored information}*
Herbrand-Interpretation *f*   free interpretation
herleitbar   deductive
Herleitung *f*   deduction
hervorrufen   evoke/to *{response}*
Heuristik *f*   heuristics
heuristisch   heuristic
  heuristische Erkennungsmethode *f*   heuristic recognition method
  heuristische Methode *f*   heuristic method
  heuristische Programmierung *f*   heuristic programming
  heuristische Regel *f*   heuristic rule
  heuristische Suche *f*   heuristic search
  heuristisches Modell *n*   heuristic model
  heuristisches Programm *n*   heuristic routine
  heuristisches Wissen *n*   experiential knowledge, heuristic knowledge, surface knowledge
Hierarchie *f*   hierarchy *{ranked series}*
Hierarchieebene *f*   hierarchical level, hierarchy level
Hierarchiestruktur *f*   hierarchical structure
Hierarchiesystem *n*   hierarchical system
hierarchisch   hierarchical
  hierarchisch geordnet   graded, ranked
  hierarchische Klassifikation *f*   hierarchical classification
  hierarchische Ordnung *f*   hierarchical order
  hierarchische Steuerungsstruktur *f*   control hierarchy
  hierarchische Struktur *f*   hierarchical structure
  hierarchisches Computernetz *n*   hierarchical computer network
  hierarchisches Computersystem *n*   hierarchical computer system
  hierarchisches Netz *n*   hierarchical network
  hierarchisches Planen *n*   hierarchical planning
  hierarchisches System *n*   hierarchical system
Hieroglyphe *f*   hieroglyph
Hilbert-Raum *m*   Hilbert space
Hilfsmittel *n*   tool
Hilfsraummethode *f*   subspace method *{a pattern-recognition method}*
Hilfssatz *m*   lemma
Hilfsvariable *f*   auxiliary variable
Hillock *n*   hillock
hinreichende Bedingung *f*   sufficient condition
hintereinandergeschaltet   connected in series
Hinterglied *n*   postcedent *{logic}*

Hippocampus m hippocampus
{extension of the cortex buried in the centre of the brain}
Hit m hit *{result of search}*
hochentwickelt sophisticated
hoch high
hochorganisiertes System n high-level system
Höchstwert m maximum
höhere Computersprache f high-order language
Hologramm n hologram
homolog homologous
homologe Gruppe f homology group
homomorph homomorphic
homomorphe Abbildung f homomorphic mapping
homomorphes Bild n homomorphic image
Homomorphie f homomorphy, homomorphism
Homomorphieprinzip n homomorphism theorem
Homomorphismus m homomorphism
Homöostase f homeostasis
homöostatischer Mechanismus m homeostatic mechanism
homöostatischer Prozess m homeostatic process
homöostatisches System n homeostatic system
Homöostat m homeostat
homotope Gruppe f homotopy group
Hopfield-Assoziator m Hopfield associator *{associative memory network in which incomplete input information causes the network to follow paths to a nearby energy minimum where complete information is stored}*
Horn-Klausel f Horn clause
Hornersches Schema n Horner's scheme
hyperaktiv hyperactive, excessively active
Hyperebene f hyperplane
Hyperfläche f hypersurface
Hyperfläche f der Zwangsbedingungen constraint hypersurface
Hypergraph m hypergraph
Hypermatrix f hypermatrix
Hyperraum m hyperspace
hypoaktiv hypoactive *{excessive inactive behaviour}*
Hypothese f hypothesis
hypothetische Maschine f hypothetic automaton
hypothetische Schlußfolgerung f hypothetic inference
hypothetischer Automat m hypothetic automaton
hypothetischer Syllogismus m modus ponens *{artificial latin expression}*

I

Idealfall m idealized situation
idealisieren idealize/to
idealisiertes Modell n idealized model
idealisiertes System n idealized system
Idealisierung f idealization
Idee f idea
Ideenverarbeitung f idea processing
Identifikationssystem n identification system

Identifikator *m* identifier *{attribute or set of attributes}*, name
Identifizierungsproblem *n* identification problem
Identität *f* identity
Identitätsaxiom *n* identity axiom
Identitätselement *n* identity element
Identitätsfunktion *f* identity function
Identitätsgatter *n* identity circuit
Identitätsglied *n* identity circuit
Identitätsglied *n* **mit zwei Eingängen** two-input identity circuit
Identitätsmatrix *f* identity matrix
Identitätsrelation *f* relationship of identity
Identitätssatz *m* identity rule, sentence of identity
Identitätsschaltung *f* identity circuit
Imitationsspiel *n* imitation game
imitieren imitate/to
IMMER ALWAYS *{quantor}*
Implementierungsumgebung *f* implementation environment
Implikation *f* implication, inclusion
implizieren imply/to
indeterminierter Automat *m* indetermined automaton
Indeterminiertheit *f* indeterminacy
indeterministisches System *n* indeterministic system
indifferente Situation *f* indifferent situation
Indifferenz *f* indifference
Indikator *m* **eines Ereignisses** event indicator
Individualkonstante *f* individual constant *{predicate logic}*
Individuenbereich *m* domain of interpretation
Individuenvariable *f* individual variable *{predicate logic}*
Individuum *n* individual *{indivisable element}*
indiziert indexed
Indizierung *f* indexing
Induktionsschluß *m* inductive conclusion
Induktionssystem *n* induction system, example-driven system *{expert system with examples as knowledge base}*
induktiv inductive
induktiv definierte Menge *f* inductive defined set
induktive Fehleranalyse *f* inductive fault analysis
induktive Schlußfolgerung *f* inductive inference, inductive conclusion
induktiver Beweis *m* proof by induction
induktiver Schluß *m* inductive conclusion, induction
induktives Lernen *n* inductive learning
induktives Shell *n* inductive shell
industrieller Roboter *m* industrial robot
Industrieroboter *m* handling machine (robot), industrial robot
Industrieroboter *m* **zum Farbspritzen** spray-painting robot
Industrieroboter *m* **zur Werkstückhandhabung** parts handling machine (robot)
Inferenzmaschine *f* inference engine
Inferenzmechanismus *m* inference engine
Inferenzregel *f* inference rule
Infimum *n* infimum
infiniter Graph *m* infinite graph

Infix-Notation *f* algebraic notation
Information *f* information
informationell informational
informationelle Redundanz *f* informational redundancy
Informationsaufnahme *f* information acquisition, information pick-up
Informationsaustausch *m* dialog[ue], information interchange
Informationsbanksteuerung *f* knowledge-based control
Informationsblock *m* information block
Informationsdarstellung *f* information representation
Informationsdefizit *n* information lack
Informationsdichte *f* information density
Informationsempfänger *m* information receiver
Informationsentropie *f* average information content
Informationsfluß *m* information flow
Informationsgehalt *m* information content
Informationsgehalt *m* **eines Merkmals** information content of a feature
Informationsgewinnung *f* information gathering
Informationsinhalt *m* information content
Informationskette *f* information chain
Informationsmangel *m* information lack
Informationsmaß *n* **einer Sprache** information content of a language
Informationsmenge *f* amount of information
Informationsnutzung *f* information utilization
Informationsquelle *f* information source
Informationsraum *m* information space
Informationsrecherchesystem *n* information retrieval system
Informationsreduktion *f* information reduction
Informationsredundanz *f* informational redundancy
Informationssenke *f* information drain, information sink
Informationsstrom *m* information flow
Informationsstruktur *f* information structure
Informationsstruktur *f* **einer Computerarchitektur** machine information structure
Informationssystem *n* information system
Informationstheorie *f* information theory
Informationsumformungsprozess *m* information conversion process
Informationsverdichtung *f* information reduction
Informationsvorrat *m* information supply
Informationswiedergewinnung *f* information retrieval
informelle Schlußfolgerung *f* natural deduction
Inhalt *m* content *{e.g. point set}*
inhaltsadressierbar content-addressable
Inhaltsanalyse *f* content analysis
inhärent inherent

Inhibition *f*   inhibition; exclusion
inhomogen   non-homogeneous
inhomogener Fluß *m* **im Transportnetz**   non-homogeneous flow in the transport network
inhomogenes Transportnetzproblem *n*   non-homogeneous transport network problem
initial   initial
initialer Automat *m*   fixed-initial state automaton, initial automaton
Initialmenge *f*   initial set
Injektion *f*   injection, injective function, one-to-one function, invertible function *{logic}*
Inklusion *f*   inclusion
Inklusionsrelation *f*   inclusion relation
inklusives ODER *n*   inclusive OR
inklusives ODER-Element *n*   inclusive OR element
inkrementaler Integrierer *m*   decision (saturating) integrator
innere Ausgabe *f*   internal output *{system theory}*
innere Eingabe *f*   internal input *{system theory}*
innere Instabilität *f*   intrinsic instability
innere Straffunktion *f*   internal penalty function
innerer Punkt *m*   interior point *{topology}*
innerlich   internal
innerlich stabile Menge *f*   internal-stable set
innewohnend   inherent
Innovation *f*   innovation
Innovationsdarstellung *f*   innovation representation
Innovationsstrategien *fpl*   innovation strategies
innovative Computerarchitektur *f*   innovative computer architecture
Input *m*   input
Inputneuron *n*   input neuron *{receives data from the outside world}*
instabil   astable, instable
instabiler Zustand *m*   instable state
instabiles System *n*   unstable system
Instabilität *f*   instabilitiy
Instantiation *f*   instantiation
Instanziierung *f*   instantiation
instationärer Zufallsprozeß *m*   non-stationary random process
Instruktionsaufbau *m*   instruction structure
Instruktionsmenge *f*   instruction set
Instruktionsstruktur *f*   instruction structure
Integrationsweg *m*   path of integration
intelligent   intelligent, smart
intelligente Arbeitshilfe *f*   intelligent job aid
intelligente computerunterstützte Unterweisung *f*   intelligent computer-aided instruction
intelligente Karte *f*   smart card
intelligente Leiterplatte *f*   smart card
intelligente Zeichenerkennung *f*   intelligent character recognition
intelligenter Arbeitsplatz *m*   intelligent workstation, professional workstation
intelligenter Roboter *m*   intelligent robot
intelligentes Assistenzprogramm *n*   intelligent assistant program

intelligentes Gerät *n*   intelligent device
intelligentes Lehrsystem *n*   intelligent tutoring system
intelligentes Meßgerät (Meßinstrument) *n*   intelligent measuring instrument
intelligentes Terminal *n*   smart terminal
intelligentes Wissensbasissystem *n*   intelligent knowledge-based system
Intelligenz *f*   intelligence
Intelligenzverstärker *m*   intelligence amplifier
intensionale Logik *f*   intensional logic
Interaktion *f* **des Menschen mit dem Computer**   interaction of man and computer
interaktiv   interactive
 interaktive Anwendung *f*   interactive application
 interaktive Graphik *f*   interactive graphics
 interaktive Umgebung *f*   interactive environment
Interdependenz *f*   interdependence
Interessenbereich *m*   domain
interessierender Problembereich *m*   domain
Interface *n*   interface
 Interface *n* **für natürliche Sprache**   natural-language-interface
intern   internal
internationales Phonetikalphabet *n*   international phonetic alphabet
interne Darstellung *f*   internal representation
interne Prozedur *f*   internal procedure
internes Modell *n*   internal model

Interpolationsnetzwerk *n*   interpolative network {*gathers a clearer set of data from a noisy set*}
Interpretation *f*   interpretation
interpretative Maschine *f*   interpretative machine
interpretatives Übersetzungsprogramm *n*   interpretative translation program
Interpreter *m*   interpreter
interpretieren   interpret/to
Interpretierer *m*   interpreter
Invariantenbildung *f*   concept formation
Invarianz *f*   invariance
Invarianztheorem *n*   invariance theorem
invers   inverse
 inverses Element *n*   inverse element
Inversion *f*   inversion
Inverter *m*   invert gate, negator, NOT element
invertierbar   invertible
 invertierbare Funktion *f*   invertible function
Invertierbarkeit *f*   invertibility
invertieren   invert/to, negate/to
invertierendes Gatter *n*   invert gate
Inzidenzfunktion *f*   incidence function
Irreflexivität *f*   irreflexivity, reflexivity
irrelevant   irrelevant; ambiguous
 irrelevante Information *f*   irrelevant information
Irrelevanz *f*   irrelevance; ambiguity
Irrfahrt *f*   random walk
Irrtum *m*   error
isoliertes System *n*   [absolutely] autonomous system

isomorph isomorphous
isomorphe Gruppe *f* isomorphic group
Isomorphieproblem *n* **dynamischer Systeme** isomorphism problem of dynamic systems
Isomorphismus *m* isomorphism, bijective homomorphism
Isomorphismus *m* **von Automaten** isomorphism of automata
isoperimetrisches Problem *n* isoperimetric problem
Istwert *m* actual value
Istzustand *m* actual state
Iterationsindex *m* cycle index

J

Ja/Nein-Antwort *f* yes-or-no answer
JEDOCH-NICHT-Gatter *n* inhibitory gate
Job *m* job
Job-Wartschlange *f* job queue
Junktor *m* propositional connective, Food's algorithm, functor
Justage *f* adjustment
justiert adjusted
Justierung *f* adjustment

K

K-Plan *m* Karnaugh map
Kalkül *m* calculus, deduction system
Kanal *m* trunk
Karnaugh-Plan *m* Karnaugh map
Karte *f* map
kartesische Koordinaten *fpl* Cartesian coordinates
kartesischer Raum *m* Cartesian space
kausal [bedingt] deterministic
Keim *m* seed
Kenner *m* identifier
Kenngröße *f* [characteristic] parameter
Kenngrößen *fpl* **von Zufallsvariablen** characteristic values of random variables
Kennlinie *f* characteristic [curve]
Kenntnis gewinnen gain knowledge/to
Kenntnisstand *m* knowledge level
Kennung *f* identifier
Kennwert *m* characteristic value
Kennzeichen *n* mark
Kennzeichnung *f* marking
Kennziffer *f* characteristic
Kern *m* nucleus {*pl. nuclei, a collection of nerve cells in the gray matter of the brain*}
Kette *f* [bit] chain
Kette *f* **mit einfachen Gelenken** simple-jointed chain
Keytext *m* key phrase, key text
KI artificial intelligence, AI
KI-System *n* **mit neuronalem Netzwerk** artificial neural system
KI-Technik *f* knowledge engineering
Kinematik *f* **des starren Körpers** rigid-body kinematic
Klartext *m* decoded text
Klasse class
Klasseneinteilung *f* class division
Klassenelement *n* class element
Klassenhierarchie *f* class hierarchy
Klassenzuweisung *f* class assignment
Klassifikator *m* classifier

klassifizieren classify/to
klassische Konditionierung *f* classical conditioning *{a form of learning in which an unconditioned stimulus is presented a conditioned one and vice versa}*
Klausel *f* clause
Klauselmenge *f* set of clauses
kleines Wissenssystem *n* small knowledge system
Kleinstwert *m* minimum value
Kleinteile *npl* hardware
Klötzchenwelt *f* block world
Knapsackproblem *n* knapsack problem
Knoten *m* node *{graph}*
  Knoten *m* **des Suchbaumes** choice point
  Knoten betreffend nodal
Knotenausgang *m* node output
Knotenbasis *f* node base
knotenbewerteter Graph *m* node-evaluated graph
Knotenbewertung *f* node evaluation
Knoteneingang *m* node input
Knotengleichung *f* node equation
Knotenmenge *f* node set
knotenorientierter Netzplan *m* node-oriented graph
Knotenpotentialmethode *f* node potential method
Knotenpunkt *m* branch point, nodal point, node
Knotenpunktvektor *m* nodal vector
Knotenspaltung *f* node splitting
Knotenvariable *f* node variable
Knotenverschmelzung *f* node fusion
Kognition *f* cognition
Koinzidenz *f* coincidence, logic AND
Kombination *f* combination

Kombinatorik *f* combinatorics
kombinatorische Explosion *f* combinatorial explosion
Kommentar *m* remark
Kommunikation *f* **in natürlicher Sprache** natural-language communication
Kommunikation *f* **zwischen System und Umwelt** system-environment communication
kommutative Gruppe *f* Abelian group
Kommutativität *f* commutability
Kompatibilität *f* compatibility
kompilieren compile/to
kompiliertes Wissen *n* compiled knowledge
komplex complex
Komplexität *f* complexity
kompliziert sophisticated
Kompliziertheit *f* complexity
Kondition *f* condition
konditioniertes Lernen *n* conditioned learning
Konfidenzfaktor *m* certainty factor, confidence factor
Konfiguration *f* configuration, pattern
Konfiguration *f* **einer programmierbaren Maschine** internal configuration
Konfiguration *f* **eines Datenverarbeitungssystems** configuration
konfigurierbar configurable
Konfigurierbarkeit *f* configurability
konfigurieren configurate/to
Konflikt *m* conflict
Konflikt *m* **bei Parallelarbeit** parallel-work conflict
Konfliktauflösung *f* conflict resolution

Konfliktlösung *f* conflict resolution
Konfliktmenge *f* conflict set
Konfliktsituation *f* conflict situation, conflict case
konform conformal
  konforme Abbildung *f* conformal mapping
  konformes Glied *n* conformal element
Kongruenz *f* congruence
Kongruenzrelation *f* congruence relation
Kongruenzsystem *n* congruence system
Königsberger Brückenproblem *n* problem of the seven bridges of Koenigsberg
konjugierter Gradient *m* conjugate gradient
  konjugierter Pol *m* conjugate pole
Konjunktion *f* conjunction, AND operation
  konjunktive Normalform *f* conjunctive normal form
konkave Funktion *f* concave function
Konklusion *f* conclusion
Konkordanz *f* concordance
konkreter Automat *m* concrete automaton
konkurrente Arbeitsweise *f* concurrent mode
konkurrierend competitive
  konkurrierende Anforderung *f* contention
Konnektionismus *m* connectionism
Konsequenz *f* consequence
Konsequenzzeichen *n* consequence symbol
konservatives System *n* conservative system

konsistent consistent
Konsistenz *f* consistency, consistence
Konsistenzproblem *n* consistency problem
konstant constant, stable, steady
Konstante *f* constant
  konstante Funktion *f* constant function
Konstantenspeicher *m* glossary
Konstantsummenbedingung *f* constant-sum condition
Konstantsummenspiel *n* constant-sum game
konstruieren construct/to; design/to
konstruktive Analysis *f* constructive analysis
konstruktive Logik *f* constructive logic
Konsultation *f* consultation
Konsultationstreiber *m* consultation driver
Kontaktalgebra *f* contact algebra
Kontext *m* context
Kontextbaum *m* context tree, object tree
Kontextelement *n* context element
kontextfreie Grammatik *f* context-free grammar
  kontextfreie Sprache *f* context-free language
kontextsensitive Grammatik *f* context-sensitive grammar
Kontingenz *f* contingency
kontinuierlich continuous
  kontinuierliche Größe *f* continuous variable
  kontinuierliche Simulation *f* continuous simulation
  kontinuierlicher Zusammenhang *m* continuity

kontinuierliches Spiel *n* continuous game
Kontinuität *f* continuity
Kontinuitätsbedingung *f* continuity condition
Kontinuitätsgleichung *f* continuity equation
Kontinuumproblem *n* continuum problem
Kontradiktion *f* contradiction
kontradiktorisch contradictory
kontradiktorischer Ausdruck *m* contradictory form
Kontraposition *f* contraposition
Kontravalenz *f* contravalence, exclusive OR
kontravariant contravariant
Kontravarianz *f* contravariance
konventionelle Maschinenebene *f* conventional machine level
konventionelle Maschinensprache *f* conventional machine language
konventioneller Computer *m* conventional computer
Konvergenz *f* convergence
Konvergenzbereich *m* convergent domain, region of convergence
Konvergenzbeweis *m* convergence proof
Konvergenzgeschwindigkeit *f* convergence rate
Konvergenzkriterium *n* convergence attribute, convergence criterion
konvergierend convergent
Konvertierung *f* conversion; translation
konvex convex
    konvexe Funktion *f* convex function

konvexe Menge *f* convex set
konvexe Optimierung *f* convex optimization
konvexes Spiel *n* convex game
Konzentration *f* concentration
konzentrieren concentrate/to {e.g. parameters}
konzentriert concentrated
Konzept *n* concept; design
Konzeptbeschreibung *f* concept description
Konzeptbildung *f* concept formation
konzeptionell conceptual
konzeptionelle Datenstruktur *f* conceptional data structure
konzeptionelles Schema *n* conceptual schema
Kooperation *f* cooperation
kooperative Spieltheorie *f* cooperative game theory
kooperatives Mehrpersonenspiel *n* cooperative multiperson game
kooperatives Spiel *n* cooperative game
Koordinatenkreuz *n* coordinate system
Koordinatensystem *n* coordinate system
Koordination *f* coordination
Koordinationsproblem *n* coordination problem
Koordinator *m* coordinator
koordinieren coordinate/to
Koordinierungsstrategie *f* coordination strategy
Kopiator *m* duplicator
kopieren copy/to
Kopierer *m* duplicator
Koppelbeziehung *f* interrelation
Kopplung *f* coupling

Kopplungsbeziehung *f* coupling relation
Kopplungsrelation *f* coupling relation
Koprozessor *m* coprocessor
Körper *m* field *{Mathematik}*
Körperaxiom *n* field axiom
Körpererweiterung *f* field extension
Korrektheit *f* correctness, soundness
Korrektor *m* corrector
Korrektur *f* höherer Ordnung higher-order correction
Korrekturmöglichkeit *f* correcting capability
Korrelation *f* correlation
Korrelationscomputer *m* correlation computer
Korrelator *m* correlator
korrelieren correlate/to *{statistics}*
korreliert correlated
Korrespondenz *f* correspondence *{set theory}*
korrigieren zwecks Aktualisierung update/to
Korruption *f* corruption *{theory of games}*
Kovarianzmatrix *f* covariance matrix
Kozustandsgleichung *f* adjoint system equation
kreativ creative
  kreativer Prozeß *m* creative process
kreisen cycle/to *{simplex algorithm}*
Kreuzmenge *f* set product
Kriterium *n* criterion
  Kriterium *n* **der Autonomie** criterion of non-interaction
kritisch critical
  kritische Aktivität *f* critical activity

kritische Stabilität *f* critical stability
kritischer Weg *m* critical path
Kugelgelenkroboter *m* spherically jointed robot
Kugelkoordinatenroboter *m* spherical coordinate robot
Kundenstrom *m* arrival stream
Kunsthand *f* artificial hand
künstlerisch creative
künstlich artificial
  künstliche Abbildung *f* artificial mapping *{transformation of the inputs into an internal representation in the network}*
  künstliche Hand *f* artificial hand
  künstliche Intelligenz *f* artificial intelligence, AI
  künstliche Roboterintelligenz *f* robotic artificial intelligence
  künstliche Sprache *f* artificial language
  künstliches Gehirn *n* artificial brain
  künstliches System *n* artificial system
Kurzdarstellung *f* abstract
kürzen simplify/to, minimize/to *{switching function}*
Kürzungsmethode *f* simplifying method
Kurzzeitgedächtnis *n* dynamic knowledge base, short-term memory, working memory
KWOC-Register *n* keyword-out-of-context register
Kybernetik *f* cybernetics
Kybernetiker *m* cybernetician
kybernetisch cybernetic[al]
  kybernetische Abstraktion *f* cybernetic[al] abstraction

kybernetische Maschine *f* cybernetic[al] machine
kybernetische Maus *f* cybernetic[al] mouse
kybernetische Modellierung *f* cybernetic[al] modelling
kybernetische Schildkröte *f* cybernetic[al] turtle
kybernetischer Simulator *m* cybernetic[al] simulator
kybernetisches Modell *n* cybernetic[al] model
kybernetisches Spielzeug *n* cybernetic[al] toy
kybernetisches System *n* cybernetic[al] system
kybernetisches Tier *n* cybernetic[al] animal

L

Labyrinthautomat *m* labyrinth automaton
Lacksprühroboter *m* paint-spraying robot
Ladeproblem *n* cargo-loading problem
Lagerhaltung *f* stocking
Lagerhaltungskosten *pl* stock-keeping cost {theory of stock-keeping}
Lagerhaltungsmodell *n* inventory model, stock-keeping model
Lagerhaltungsproblem *n* stock-keeping problem
Lagerhaltungssystem *n* stock-keeping system
Lagerhaltungstheorie *f* stock-keeping theory
Lagerhaltungsüberwachung *f* stock control
Lagerkosten *pl* stock-keeping cost
Landkarte *f* map
Langzeitgedächtnis *n* long-term memory
laterale Rückführung *f* lateral feedback {feedback in a multilayered network such that an input of a neuron depends on the output of other neurons of the same layer}
laufen run/to {e.g. program}
laufend current
laufende Information *f* current information
laufender Betrieb *m* running mode {mode after the network has learned, weight are constant und being used to make decisions, associations etc.}
laufender Zustand *m* current state
Laufzeitsystem *n* [run-]time system
Leap-frog-Struktur *f* leapfrog structure
leer empty
leere Liste *f* empty list {list containing zero elements}
leere Menge *f* empty set, null set
leere Schale *f* empty shell {expert system}
leerer Graph *m* empty graph
leeres Shell *n* empty shell {expert system}
Lehrautomat *m* teaching automaton (computer, machine)
lehren teach/to, instruct/to {learning automaton}
lehren/den Roboter eine neue Arbeitsaufgabe teach the robot a new job/to
Lehrhilfe *f* teaching tool
Lehrmaschine *f* teaching automaton (computer, machine)

Lehrmuster *n* teaching pattern
Lehrprogramm *n* teaching program; educational Software
Lehrprogrammentwickler *m* instructional designer
Lehrschritt *m* teaching step
Lehrsystem *n* teaching system, tutor system
leisten perform/to
Leitcomputer *m* master computer
leiten control/to
Leitgerät *n* master
Leithandhabungsautomat *m* master manipulator
Leitmanipulator *m* master manipulator
Leitungssystem *n* management system
Leitungswissenschaft *f* management science
Leitweg *m* routing
Lemma *n* lemma
lenken control/to
Lernalgorithmus *m* learning algorithm
Lernelement *n* learning element
lernen learn/to
  Lernen *n* learning *{a neural network modifies its weights in response to external input}*
  Lernen *n* **aus Beispielen** concept acquisition
  Lernen *n* **des Computers** machine learning
  Lernen *n* **durch bedingte Zuordnung** learning by conditional allocation
  Lernen *n* **durch Begreifen** *s.* Lernen *n* durch Erfassen
  Lernen *n* **durch Belehrung** learning by instruction
  Lernen *n* **durch Erfassen** learning by understanding, learning by comprehending
  Lernen *n* **durch Erfolg** learning by success
  Lernen *n* **durch Erklärung** explanation-based learning
  Lernen *n* **durch Nachahmung** learning by copying, learning by imitating
  Lernen *n* **durch Optimieren** learning by optimizing
  Lernen *n* **durch Speichern** learning by storing
  Lernen *n* **durch Verrichten** learning by doing
  Lernen *n* **in nichtstationärer Umgebung** learning without steady-state conditions, non-linear learning
  Lernen *n* **in stationärer Umgebung** learning with steady-state conditions
  Lernen *n* **unter dem Einfluß von Störgrößen** learning without steady-state conditions, non-linear learning
  Lernen *n* **zweiter Ordnung** second order learning
lernend learning; self-adaptive
lernende Maschine *f* learning automaton, learning machine
lernende Regelung *f* learning control
lernender Automat *m* learning automaton, learning machine
lernender Computer *m* learning computer, self-learning computer
lernender Klassifikator *m* learning classifier
lernendes System *n* learning system, self-learning system

lernfähig learnable, adaptive, trainable
lernfähige Logik *f* adaptive logic
lernfähiger Automat *m* trainable system (automaton)
lernfähiges Element *n* learning element
Lernfähigkeit *f* adaptation, learnability
Lernklassifikator *m* learning classifier
Lernmatrix *f* learning matrix
Lernmechanismus *m* learning mechanism
Lernmenge *f* learning set
Lernmodell *n* learning model
Lernmodelltechnik *f* learning-model technique
Lernmuster *n* learning pattern
Lernphase *f* learning phase
Lernprogramm *n* learning program
Lernprozeß *m* learning process
Lernrate *f* learning rate {*factor to scale all corrections while learning to improve the convergence speed of the network*}
Lernregel *f* learning rule {*a Hebb's rule that tells the network how to modify the values of the connection weights*}
Lernschema *n* learning scheme
Lernsoftware *f* instructional software, teachware
Lernstruktur *f* learning structure
Lernsystem *n* learning system, self-learning system
Lerntheorie *f* learning theory
Lernverbindung *f* learning connection
Lernverhalten *n* learning behaviour
Lernvermögen *n* memorization

lesbar legible
Leseautomat *m* **nach dem Korrelationsprinzip** reading automaton according to correlation principle
Lesen *n* reading; sensing
letzter Knoten *m* terminal (leaf) node
lexikalischer Analysator *m* lexical analyzer
lexikographischer Code *m* lexicographic code
Likelihood *f* likelihood
linear linear
linear unabhängig [voneinander] orthogonal
linearer Vektorraum *m* linear vector space
lineares Lernen *n* learning with steady-state conditions, linear learning
linearisiertes Modell *n* linearized model
Linearitätsbereich *m* linearity range
Linksverschiebung *f* shift left
Linksystem *n* link system {*queueing theory*}
LISP LISP {*a programming language preferably for expert systems*}
Liste *f* list
Liste erstellen retrieve/to
Listendarstellung *f* list representation
Listenverarbeitung *f* list processing
Literal *n* literal
Logik *f* logic
Logik *f* **der programmierbaren Felder** programmable array logic
Logik *f* **mit drei Zuständen** three-state logic
Logikanalysator *m* logic analyzer, logic circuit analyzer

Logikanalyse *f* logic analysis
Logikanordnung *f* logic array
Logikdiagramm *n* logic diagram
Logikelement *n* logic element
Logiker *m* logician
Logikkalkül *m* deduction system
Logikoperation *f* logic operation
Logikplan *m* logic diagram
Logikschaltung *f* logic circuit
Logikvariable *f* logic variable
logisch logical, Boolean
 logisch fortlaufend organisierter Rechenautomat *m* arbitrary-sequence computer
 logisch gültig [universally] valid, asserted
 logisch-mathematische Sprache *f* logical-mathematical language
 logisch-mathematisches Kalkül *m* logical-mathematical calculus
 logisch schließen derive by reasoning/to, infer/to
 logische Adresse *f* logical adress
 logische Äquivalenz *f* equivalence
 logische Aussage *f* proposition, symbolic sentence
 logische Beweisführung *f* reasoning
 logische Elementarfunktion *f* logical elementary function
 logische Elementarverknüpfung *f* elementary logical connection
 logische Entscheidung *f* logical decision
 logische Folgerung *f* consequence, logical inference
 logische Gleichung *f* Boolean equation, logical equation
 logische Gleichwertigkeit *f* logical equivalence
 logische Grundfunktion *f* basic logical function
 logische Matrix *f* logical matrix
 logische Multiplikation *f* conjunction
 logische Null *f* at-rest signal
 logische Operation *f* logical operation
 logische Schaltung *f* logical (switching) circuit
 logische Semantik *f* logical semantics
 logische Summe *f* disjunction, logical sum
 logische UND-Verknüpfung *f* coincidence, logical AND
 logische UND-Verknüpfung *f* logical AND
 logische Verbindungsstelle *f* logical node
 logische Verknüpfung *f* logical [inter]connection, logic circuit
 logische Verknüpfungsschaltung *f* logical circuitry; gate
 logischer Ausdruck *m* logical (Boolean) expression, proposition
 logischer Automat *m* logical automaton, logical machine
 logischer Befehl *m* logical instruction
 logischer Folgerungsschritt *m* pro Sekunde logical inference step per second
 logischer Grundbaustein *m* basic logical module
 logischer Knoten *m* logical node
 logischer Operator *m* quantifier, logical connective
 logischer Schluß *m* logical consequent, logical inference
 logischer Vergleich *m* logical comparison

logischer Widerspruch *m* logical contradiction
logisches Diagramm *n* logical diagram
logisches Feld *n* logical array
logisches Folgeelement *n* sequential logical element
logisches Grundelement *n* basic logical element
logisches Produkt *n* conjunction
logisches Verknüpfungselement *n* logical (switching) element
logisches Verknüpfungsglied *n* gate
logisches Zeichen *n* quantifier
Logistik *f* logistics
lokale Stabilität *f* local stability
lokale Variable *f* local variable
lokales Extremum *n* local extremum
lokales Maximum *n* local maximum
lokales Minimum *n* local minimum
lokales Optimum *n* local optimum
Lokalisierung *f* localization *{occurs when a set of neurons which are close together receives a set of signals in parallel and responds as a unit}*
Lokalitätsprinzip *n* principle of locality
lösbar solvable *{e.g. system of equations}*
löschen clear/to
lösen solve/to; solute/to
Lösen *n* solving
lösen/ein Problem solve a problem/to
Lösung *f* solution
Lösung *f* **im statischen (eingeschwungenen) Zustand** steady-state solution
Lösungsalgorithmus *m* solution algorithm
Lösungsbaum *m* solution tree
Lösungsgraph *m* solution graph
Lösungsmodell *n* solution model
Lösungspfad *m* solution path *{e.g. through a search space}*
Lösungsweg *m* approach, path

M

Mächtigkeit *f* potency *{set}*
Majorität *f* majority
Majoritäts-Minoritäts-Logik *f* majority-minority logic
Majoritätselement *n* majority element
Majoritätsentscheidungsgatter *n* majority-decision gate
Majoritätsentscheidungslogik *f* majority-decision logic
Majoritätsfunktion *f* majority function
Majoritätsgatter *n* majority element (gate)
Majoritätsglied *n* majority element (gate)
Majoritätsschaltung *f* majority element
Makro-E-Netz *n* macro evaluation net
Makrostelle *f* macro location
Managementautomatisierung *f* automation of management
Mangel *m* shortage
Mangel *m* **an Wissen** lack of knowledge
Mangelkosten *pl* shortage cost *{stock-keeping}*

Manipulator *m* industrial robot, manipulator-type robot
manipulieren manipulate/to
Manipulierung *f* manipulation
Mannigfaltigkeit *f* manifold *{group theory}*
Mantel *m* **eines Expertensystems** runtime system
markieren mark/to, tag/to
Markieren *n* marking
Markierung *f* marking
Markierungsfunktion *f* marking function *{Moore automaton}*
Markierungsmodell *n* marker model *{graph technique}*
Markow-Algorithmus *m* Markov algorithm
Markow-Kette *f* Markov (Markovian) chain
Markow-Prozeß *m* Markov process
Masche *f* mesh *{of a network}*
Maschennetz *n* mesh network
Maschine *f* machine
maschinell lesbar computer-readable
  maschinelle Bilderkennung *f* computational (computer, machine) vision
  maschinelle Deduktion *f* logical inference
  maschinelle Spracherkennung *f* machine recognition of speech
  maschinelle Sprachübersetzung *f* machine translation of languages
  maschinelle Übersetzung *f* automatic (machine) translation
Maschinenadresse *f* absolute address
Maschinenbefehl *m* instruction
Maschinencode *m* absolute code
Maschinenebene *f* machine level
maschinengestützte Übersetzung *f* machine-aided translation of languages
Maschinenintelligenz *f* artificial intelligence
maschinenlesbar computer-readable
Maschinensprache *f* machine language
Maschinenübersetzung *f* machine translation
Maschinenübersetzung *f* **mit menschlicher Unterstützung** human-aided machine translation
Maschinenwörterbuch *n* machine dictionary
Massenbedienungstheorie *f* mass servicing theory, queueing theory
maßgeschneidert tailored
maßstabgetreues Modell *n* scale model
Mathematik *f* mathematics
mathematisch mathematical
  mathematische Argumentform *f* **in deduktiver Logic** modus ponens
  mathematische Darstellung *f* mathematical representation
  mathematische Energiefunktion *f* computational energy function *{a mathematical function defining the stable states of a network and the paths to them}*
  mathematische Erwartung *f* mathematical expectation
  mathematische Linguistik *f* mathematical (algebraic) linguistics
  mathematische Logik *f* mathematical logic
  mathematische Modellierung *f* mathematical modelling
  mathematische Optimierung *f* mathematical optimization

mathematische Semantik *f* mathematical semantics
mathematische Simulierung *f* mathematical simulation
mathematische Statistik *f* mathematical statistics
mathematisches Entscheidungsmodell *n* mathematical decision model
mathematisches Gleichungssystem *n* array of mathematical equations
mathematisches Sprachmodell *n* mathematical language model
Mathematisierung *f* mathematization
Matrix *f* matrix; array
  Matrix *f* **der partiellen Wirkung** matrix of partial effect
Matrixdarstellung *f* matrix representation
Matrixminimummethode *f* minimum matrix procedure
Matrixschreibweise *f* matrix notation
Matrixspeicher *m* array store
Matrixspiel *n* matrix game, two-person zero-sum game
Maximalfehler *m* maximum error
Maximalflußproblem *n* maximal flow problem
Maximalkettensatz *m* maximum chain theorem *{set theory}*
Maximalprinzip *n* maximum principle
Maximalstromproblem *n* maximum-flow problem *{graph theory}*
Maximalwertsuche *f* maximum-value search
Maximaxprinzip *n* maximax principle
maximierbar maximable

Maximierbarkeit *f* maximability
maximieren maximize/to
Maximierung *f* maximization
  Maximierung *f* **des Gewinns** gain maximization
Maximierungsalgorithmus *m* maximization algorithm
Maximierungsfunktion *f* maximizing function
Maximierungskriterium *n* maximization criterion
Maximierungsproblem *n* maximization problem
Maximierungsproblem *n* **mit Randbedingungen** constrained maximization problem
Maximierungsproblem *n* **ohne Randbedingungen** unconstrained maximization problem
Maximinkriterium *n* maximin criterion
Maximinprinzip *n* maximin principle
Maximinstrategie *f* maximin strategy
Maximintheorem *n* maximin theorem
Maximum *n* maximum
Maximum-Likelihood-Prinzip *n* principle of least squares
Maximumprinzip *n* maximum principle
Maxterm *m* elementary alternative, maxterm
Mealyscher Automat *m* Mealy automaton *{sequential circuitry}*
Means-end-Analyse *f* means-end analysis
Mechanismus *m* mechanism
Mehrcomputersystem *n* distributed-intelligence computer system

mehrdeutig ambiguous, non-unique
mehrdeutige Abbildung *f* non-unique mapping
Mehrdeutigkeit *f* ambiguity; equivocation
mehrdimensionale Zufallsgröße *f* random vector
Mehrebenensystem *n* multilevel system
mehrfach verzweigtes Netz[werk] *n* multibranched network
mehrfache Vererbung *f* multiple inheritance
Mehrfachkanten *fpl* parallel edges
Mehrfelderbildschirm *m* split screen
Mehrheit *f* majority
Mehrheitselement *n* majority element
Mehrheitsgatter *n* majority element
Mehrheitsglied *n* majority element
Mehrheitslogik *f* majority logic
Mehrheitsschaltung *f* majority element
Mehrheitssystem *n* majority system
Mehrpersonenspiel *n* multiperson game
mehrschichtig multilayered {*neuronal network*}
Mehrstufenentscheidungsprozeß *m* multistage decision process
mehrwertige Logik *f* many-valued logic
mehrwertiges Attribut *n* multi-valued attribute
Mehrwertigkeit *f* polyvalence
mehrzweigiges Modell *n* multibranched model
Membran *f* membrane {*cellular tissue sorrounding the neuron*

cell body}
Menge *f* amount; set {*set theory*}
Menge *f* **der Mittel** means set
Menge *f* **erster Ordnung (Stufe)** first-level set, first-order set, fixed level set
Menge *f* **zweiter Ordnung (Stufe)** second-level set
Mengenalgebra *f* set algebra
Mengenbildungsprinzip *n* set-construction principle
Mengenfamilie *f* set family, set system
Mengenfunktion *f* set function
Mengenlehre *f* set theory
Mengenoperation *f* set operation
Mengenpotenz *f* set power
Mengenprodukt *n* set product
Mengensymbol *n* set symbol
Mengensystem *n* set family, set system
mengentheoretische Topologie *f* set-theoretical topology, general topology
Mensch *m* man; human
Mensch *m* **als Bediener (Operateur)** human operator
Mensch-Maschine-Dialog *m* man-machine dialogue
Mensch-Maschine-Interface *n* man-machine interface
Mensch-Maschine-Kommunikation *f* man-machine communication
Mensch-Maschine-Kopplung *f* man-machine interaction
Mensch-Maschine-Schnittstelle *f* man-machine interface
Mensch-Maschine-System *n* man-machine system
menschenähnlicher Finger *m* humanlike finger {*part of robot*}

menschliche Informationsverarbeitung *f* human information processing
menschliche Intelligenz *f* human intelligence
menschliches Gehirn *n* human brain
Merkmal *n* characteristic, feature
Merkmalsanalyse *f* feature analysis
Merkmalsfilter *n* feature filter
Merkmalsgewinnung *f* bei Bildern figure extraction
Merkmalspaar *n* pair of characteristics
Merkmalsraum *m* space of characteristics
Merkmalsselektion *f* feature selection
Merkmalsvektor *m* feature vector
Merkmalswert *m* characteristic value
Meßdatenanalyse *f* analysis of data
Metacompiler *m* metacompiler
Metalogik *f* metalogic
Metamathematik *f* metamathematics
Metaregel *f* metarule {superimposed inference rule}
Metaspiel *n* metagame
Metasprache *f* metalanguage
metastabil metastable
Metasymbol *n* metasymbol
Metatheorie *f* metatheory
Metawissen *n* metaknowledge
Methode *f* method, approach
  Methode *f* der doppelten Gradienten dual-gradient method
  Methode *f* der kleinen Variationen small-variations method
  Methode *f* der konjugierten Gradienten conjugate-gradient method
  Methode *f* der konzeptuellen Abhängigkeit *f* conceptual dependency method
  Methode *f* der optimal brauchbaren Richtung optimum feasible-direction method {gradient method}
  Methode *f* der schrittweisen Näherung cut-and-try procedure
  Methode *f* der zufälligen Suche random search method
  Methode *f* der zulässigen Richtungen feasible-directions method {non-linear optimization}
  Methode *f* des Hypothetisierens und Testens top-down method
  Methode *f* des kleinsten Parameters smallest-parameter perturbation method
  Methode *f* des kritischen Weges critical path method
  Methode *f* des kürzesten Weges shortest-path (-route) method
  Methode *f* des Lernens durch Erfolg trial-and-error method
  Methode *f* des Scheinspiels fictitious-game method {linear optimization}
  Methode *f* des steilsten Abstiegs hill-climbing method, method of steepest descent
  Methode *f* des steilsten Aufstiegs hill-climbing method
Mikrocomputersystem *n* mit verteilter Intelligenz distributed-intelligence microcomputer system
Mikroprozessorsystem *n* mit mehreren aneinandergereihten Mikroprozessoren sliced microprocessor system
mikrozellulare Struktur *f* microcellular structure
Mindestwert *m* minimum value

Minimalbaum *m* minimum tree
minimale disjunktive Normalform *f* minimal disjunctive normal (standard) form
minimale Logik *f* minimum logic
minimaler Automat *m* minimal automaton, minimum automaton
Minimalgerüst *n* minimum framework {network}
Minimalisierung *f* **des Merkmalsatzes** minimization of the set of features
Minimalprinzip *n* minimum principle
Minimalwert *m* minimum value
Minimax-Entscheidungsregel *f* minimax decision rule
Minimaxidentifikation *f* minimax identification
Minimaxkriterium *n* minimax criterion
Minimaxprinzip *n* minimax principle
Minimaxstrategie *f* minimax strategy
Minimaxsuche *f* minimax search
Minimaxtheorem *n* minimax theorem
minimierbar minimable
Minimierbarkeit *f* minimability
minimieren minimize/to, simplify/to {switching function}
Minimierung *f* minimization
Minimierung *f* **der Anzahl der Zustände eines Automaten** state number minimization of an automaton
Minimierungsalgorithmus *m* minimization algorithm
Minimierungsfunktion *f* minimizing function
Minimierungskriterium *n* minimization criterion
Minimierungsproblem *n* minimization problem
Minimierungsproblem *n* **mit Randbedingungen** constrained minimization problem
Minimierungsproblem *n* **ohne Randbedingungen** unconstrained minimization problem
Minimierungsverfahren *n* minimization technique
Minimierungsverfahren *n* **nach Quine** minimization method due to Quine
Minimumsuchmethode *f* minimum search method
Minterm *m* elementary conjunction, minterm
mischen merge/to {combining two or more files}
mischsortieren merge/to
Mischungsproblem *n* mixing (blending) problem {optimization}
mitgeschleppter Fehler *m* inherited error
Mittel *n* average
Mittelmenge *f* mean set
mitteln average/to
Mittelwert *m* average
Mittelwert *m* **einer Zufallsfunktion** average of random function
mittlere Warteschlangenlänge *f* mean queue size
mittlerer bedingter Informationsgehalt *m* average conditional information content
mittlerer Erwartungswert *m* mean expectation
mittlerer Informationsbelag *m* average information content per symbol

Mnemonik 184

mittlerer Informationsfluß *m* average information rate per time
mittlerer Informationsgehalt *m* average information content
mittlerer Transinformationsbelag *m* [einer Nachrichtenquelle] average transinformation content per symbol
mittlerer Transinformationsfluß *m* average transinformation rate per time
mittlerer Transinformationsgehalt *m* average transinformation content
mittlerer Verzweigungsfaktor *m* average branching factor *{tree-like structure}*
Mnemonik *f* mnemonics
mnemonische Kennung *f* generic identifier
mnemonischer Kennzeichner *m* generic identifier
Mnemoschema *n* mnemonic diagram
Modalitätenlogik *f* modality logic
Modalwert *m* mode
Mode *f* mode
Modell *n* model
 Modell *n* **der Nervenzelle** model of the nervous cell
 Modell *n* **der Umgebung** environmental model
 Modell *n* **des Systems** system model
 Modell *n* **des visuellen Analysators** model of the visual analyzer
 Modell *n* **eines mehrstufigen Produktionsprozesses** model of a multistage production process
 Modell *n* **eines mehrstufigen Verteilungsprozesses** model of a multistage distribution process
 Modell *n* **mit variabler Struktur** model with variable structure
 Modell *n* **mit Wahrscheinlichkeitseinschränkung** probability-restriction model
 Modell *n* **zu erkennender Objekte** model of objects to be recognized
Modellabbildung *f* model mapping
modelladaptive Regelung *f* matching control, model-reference adaptive control
modelladaptives System *n* model-reference adaptive system
Modellbeschreibung *f* model description
Modellbewertung *f* model evaluation
Modellbildung *f* model[l]ing
Modellentwurf *m* model design
Modellextremalsystem *n* model-extremal sytem
Modellfehler *m* model error
Modellhierarchie *f* model hierarchy
modellierbar model[l]able
Modellierbarkeit *f* model[l]ability
modellieren model/to, simulate/to
Modellierung *f* model[l]ing, [mathematical] simulation
 Modellierung *f* **adaptiver Systeme** adaptive-system model[l]ing
 Modellierung *f* **der Erkennungs- und Lehrprozesse** model[l]ing of recognition and teaching processes
 Modellierung *f* **des Denkens** model[l]ing of thinking
 Modellierung *f* **des Gedächtnisses** model[l]ing of the memory
 Modellierung *f* **des Identifizierungsprozesses** identification process model[l]ing

Modellierung *f* **des Mensch-Maschine-Systems** model[l]ing of the man-machine system
Modellierung *f* **des Wahrnehmungsprozesses** model[l]ing of the perception process
Modellierung *f* **kybernetischer Systeme** cybernetic[al] system model[l]ing
Modellierung *f* **mit Bondgraph** bond graph model[l]ing
Modellierung *f* **molekularbiologischer Systeme** model[l]ing of systems of molecular biology
Modellierung *f* **realer Prozesse** real-process model[l]ing
Modellierung *f* **sensorischer Systeme** model[l]ing of sensory systems
Modellierung *f* **von Systemen** system model[l]ing
Modellierungsstrategie *f* model[l]ing strategy
Modellklasse *f* model class
Modellmethode *f* model[l]ing method
Modellnachführung *f* model updating
Modellobjekt *n* model object
Modellparameter *m* model parameter
Modellrelation *f* model relation
Modellschätzung *f* model estimation
Modellsimulierung *f* model simulation
Modellstabilität *f* model stability
Modellsteuerung *f* model-driving *{a top-down method}*
Modelltheorie *f* model theory
Modellvereinfachung *f* model simplification

Modellverhalten *n* model behaviour
Modellversuch *m* model experiment
modern advanced
modernes Bauelement *n* advanced component
modernes System *n* advanced system
Modifikation *f* modification
Modul *m* module *{smallest functional unit in the cortex}*
Modus *m* mode
Modus ponens *m* modus ponens *{artificial expression used in logic}*
Modus tollens *m* modus tollens *{artificial expression used in logic}*
mögliches Ereignis *n* possible event *{calculus of probability}*
Möglichkeit *f* possibility; likelihood
momentan current
momentaner Zustand *m* current state
Momentenmethode *f* method of moments
monadische Operation *f* monadic operation
monadische Sprache *f* monadic language
monadischer Boolescher Operator *m* monadic Boolean operator
Monoid *n* monoid *{theory of sets}*
monotone Boolesche Funktion *f* monotonic Boolean function
monotone Folgerung *f* monotonic inference (reasoning)
monotone Funktion *f* monotonic function
monotone Instabilität *f* non-oscillatory instability
monotones Schließen *n* monotonic reasoning
Monotonie *f* monotonicity

monoton monotonic
Montageroboter *m* assembly robot
Monte-Carlo-Methode *f* Monte-Carlo method, random walk method
Moore-Automat *m* Moore automaton *{sequential circuit}*
Morphismus *m* morphism
Muster *n* model, pattern; gestalt *{pattern perceived as a whole}*
Muster *n* **von Eingabedaten** input pattern
Musteranalyse *f* pattern analysis
Musterassoziationsnetz *n* pattern associator *{a network that associates two patterns}*
Musterauswertegerät *n* pattern interpreter
Mustererkennung *f* pattern recognition
Mustererkennungsgerät *n* pattern recognizer
Mustergenerator *m* pattern generator
mustergerichtetes Aufrufen *n* pattern-directed invocation
musterklassifizierendes System *n* pattern-classifying system
Musterklassifizierung *f* pattern classification
Mustervergleich *m* pattern matching
Mustervergleicher *m* pattern associator (matcher)
Musterwiederauffindung *f* pattern retrieval
Muttersprache *f* native language

N

n-wertige Logik *f* n-valued logic
nachahmen copy/to
Nachbar *m* neighbour
Nachbarneuronen *npl* neighborhood set [of neurons] *{all the neurons immediately to a particular neuron}*
Nachbedingung *f* postcondition
Nachbereich *m* range *{relation}*
nachbilden imitate/to, simulate/to, model/to
Nachbildung *f* model[l]ing, simulation
Nachereignis *n* post-event *{graph technique}*
Nachfolgeknoten *m* follow-up node
nachfolgende Aktivität *f* succeeding activity *{network technique}*
nachfolgender Knoten *m* follow-up node *{graph}*
Nachfolgerelation *f* follower relation
Nachfolgerzustand *m* follower state
Nachsatz *m* conclusion *{logic}*
Nächste-Nachbarn-Methode *f* nearest-neighbour method
Nächste-Nachbarn-Regel *f* nearest-neighbour rule
Nachweis *m* proof
Näherung *f* **erster Ordnung** first-order approximation
Näherung *f* **höherer Ordnung** higher-order approximation
Näherungsbestimmung *f* approximate determination
Näherungsgebiet *n* approximation region
Näherungsgleichung *f* approximation equation

Näherungsintervall *n* approximation interval
Näherungslösung *f* approximate solution
Näherungsmethode *f* approximation method
Näherungsmodell *n* approximation model
Näherungspolynom *n* approximation polynomial
Näherungsrechnung *f* approximation calculus
Näherungstheorie *f* approximation theory
näherungsweise approximate, approximative
Näherungswert *m* approximation value
Nahtstelle *f* **zwischen Mensch und Maschine** man-machine interface
naive Physik *f* naive physics
Name *m* identifier; name
NAND-Element *n* NAND circuit (element, gate, network)
NAND-Gatter *n* inhibitory gate, NAND circuit (element, gate, network)
NAND-Glied *n* NAND circuit (element, gate, network)
NAND-Operation *f* NAND operation, non-conjunction
NAND-Schaltung *f* NAND circuit (element, gate, network)
natürlich natural
 natürliche Instabilität *f* inherent instability
 natürliche Intelligenz *f* natural intelligence
 natürliche Sprache *f* natural language
 natürlicher Schluß *m* natural deduction
 natürliches System *n* natural system
natürlichsprachiges Interface *n* natural-language interface
Nebenbedingung *f* constraint
nebeneinander [angeordnet] adjacent
nebenläufig concurrent
Negation *f* negation, NOT operation
negative Entropie *f* negative entropy, negentropy
negative Logik *f* negative-true [logic]
Negator *m* negator, NOT element, NOT gate
Negentropie *f* negative entropy, negentropy
negieren negate/to, invert/to
negierte Disjunktion *f* non-disjunction, NOT OR operation
negierte Konjunktion *f* NAND operation, non-conjunction
negierte ODER-Funktion *f* Peirce function
Neigung *f* **zum Feuern [des empfangenden Neurons]** excitatory tendency *{tendency of a synapse or connection to cause firing of the receiving neuron}*
Neocortex *m* neocortex *{human cortex, evolutiuonary newest part of the cortex}*
Neokonnektionismus *m* fine grain parallelism
Nervenimpuls *m* nerve impulse
Nervenzelle *f* nerve cell
Netware *f* netware *{software that defines a neural network}*
Netz *n* network *{e.g. computer network, s.a. Netzwerk}*

Netz *n* **von Automaten** automata network
Netzanalysator *m* network analyzer
Netzanalyse *f* network analysis
Netzarchitektur *f* network architecture
Netzmodell *n* network model
Netzparameter *m* network parameter
Netzplan *m* network, graph
Netzplananalyse *f* network analysis
Netzplanmatrix *f* network matrix
Netzplanmodell *n* network model
Netzplantechnik *f* graph technique
Netzsteuersprache *f* network control language
Netzverwaltungssystem *n* network management system
Netzwerk *n* network *{s.a.* Netz*}*
Netzwerk *n* **mit konzentrierten Parametern** concentrated-parameter network
Netzwerkanalysator *m* network analyzer
Netzwerkanalyse *f* network analysis
Netzwerktheorie *f* network theory
Neuheitsdetektor *m* novelty detector *{an adaptive network which detects changes in the input and goes to a non-zero value if the input pattern new otherwise the output will work its way toward zero}*
neural neural *{s.a.* neuronal*}*
 Neuralcomputer *m* neural computer
neuraler Computer *m* neural computer
Neuralnetz *n* neural network
Neuristor *m* neuristor
Neurobionik *f* neurobionics
Neurocomputer *m* computer neural network, neurocomputer
Neurokybernetik *f* neurocybernetics

Neuron *n* neuron *{nerve cell or processing element in an artificial neural network}*
neuronal neuronal, neuronic
neuronaler Schaltkreis *m* neuronal network
neuronales Netz[werk] *n* brain (neuronal) network
neuronenähnliches Netz[werk] *n* neuron-like network
Neuronennetz *n* net of neurons
Neuronensimulation *f* neuron simulation
Neutralisationskondensator *m* neutralizing capacitor
Neutralzustand *m* neutral state
NICHT *n* NOT *{logic expression}*
 NICHT-Element *n* invert gate, element, negator; literal
NICHT-Funktion *f* NOT function
NICHT-Gatter *n* invert gate, NOT gate
NICHT-Glied *n* invert gate, negator, NOT element
NICHT-ODER *n* NOT OR
NICHT-ODER-Operation *f* nondisjunction, NOT OR operation
NICHT-Operation *f* NOT operation
NICHT-Schaltung *f* NOT circuit
NICHT-UND *n* NOT AND
NICHT-UND-Gatter *n* NOT AND gate
NICHT-UND-Operation *f* NAND operation, non-conjunction
NICHT-UND-Schaltung *f* inversion circuit
nicht unterscheidbar indistinguishable
Nicht-von-Neumann-Verarbeitung *f* parallel processing

nicht vorhersagbar unpredictable
nicht wechselwirkendes System *n* non-interacting (decoupled) System n
nicht weiter zerlegbare logische Aussage *f* atom
nicht weiter zerlegbarer Knoten *m* terminal node
nichtalgorithmisch non-algorithmic
nichtalgorithmischer Prozeß *m* non-algorithmic process
Nichtatom *n* literal
nichtautonomes System *n* non-autonomous system
nichtchronologisch nonchronological
nichtdeterminierter Automat *m* non-deterministic automaton
nichtdeterminiertes System *n* non-determinated system
nichtdeterministisch non-deterministic
nichtdeterministischer Algorithmus *m* non-deterministic algorithm
nichtdeterministisches Signal *n* non-deterministic signal
nichteindeutig stochastic
nichtentscheidbar undecidable
Nichtentscheidbarkeit *f* undecidability
nichtfunktionales Element *n* non-functional element
nichtgerichtet non-directional
nichtglatte Funktion *f* non-smooth function
nichthomogen non-homogeneous
nichtinitialer Automat *m* non-initial automaton
nichtinvertierende Logik *f* non-inverting logic, English logic
nichtklassische Logik *f* non-classical logic

nichtklassisches Variationsproblem *n* non-classical variational problem
nichtkonvexe Optimierung *f* non-convex optimization
nichtkooperative Spieltheorie *f* non-cooperative game theory
nichtkooperatives Spiel *n* non-cooperative game
nichtkritische Aktivität *f* non-critical activity *{network}*
nichtleere Menge *f* non-empty set
nichtlinear non-linear
nichtlineares Lernen *n* non-linear learning, learning without steady-state conditions
Nichtlinearität *f* non-linearity
nichtmonotone Logik *f* non-monotonic logic
nichtmonotones Schließen (Folgern) *n* non-monotonic inference (reasoning)
Nichtnegativitätsbedingung *f* non-negativity condition
Nichtnullsummenspiel *n* non-zero sum game
nichtoptimale Schätzung *f* non-optimum estimation
nichtplanarer Graph *m* non-planar graph
nichtprozedurale Sprache *f* non-procedural language *{e.g PROLOG}*
nichtsequentielles System *n* non-sequential system
nichtsingulärer Punkt *m* non-singular (non-critical) point
nichtstrategisches Spiel *n* non-strategic game
nichtstrukturierte Entscheidung *f* unstructured decision *{cannot be specified in terms of a set of rules}*

nichtsymmetrische Relation *f* non-symmetric relation
nichtterminaler Knoten *m* non-terminal node *{tree structure}*
nichttriviale Untergruppe *f* proper subgroup
nichtunterscheidbares Ereignis *n* non-discernible event
nichtzufällig non-random
Nickachse *f* pitch axis *{robot}*
Nicod-Funktion *f* Nicod (Peirce) function
NIE NEVER *{quantor}*
nilpotentes Ereignis *n* nilpotent event
Nilpotenz *f* nilpotency
Niveau *n* level
NN-Regel *f* nearest-neighbour rule
NOR *n* inverted OR
NOR-Funktion *f* NOR-function, Peirce function
NOR-Operation *f* disjunction, non-disjunction, NOT OR operation
NOR-Verknüpfung *f* NOR operation
Nordwesteckenregel *f* northwest corner rule
normaler Reiz *m* unconditioned stimulus
normales Verhalten *n* normal response *{to a stimulus, e.g. fear at the sight of fire}*
Normierung *f* normalization
Notationsvereinbarung *f* notation convention
notwendige Bedingung *f* necessary condition *{logic}*
NTL non-threshold logic
Nukleus *m* nucleus *{pl. nuclei, a collection of nerve cells in the gray matter of the brain}*

Null-Eins-System *n* binary system
Nullgraph *m* null graph
Nullhypothese *f* null hypothesis
Nullmatrix *f* null matrix
Nullpotentialknotenmethode *f* method of null potential nodes
Nullpunktverschiebung *f* shift in zero
Nullsignal *n* at-rest signal
Nullstelle *f* root *{mathematical function}*
Nullstellung *f* neutral position
numerische Mathematik *f* numeric[al] mathematics
numerisches Modell *n* numeric[al] model
NUR DANN, WENN ONLY THEN IF
Nutzenfunktion *f* profit function
Nutzenmatrix *f* profit matrix
Nutzinformation *f* useful information

O

Oberflächenwissen *n* experiential knowledge, heuristic knowledge, surface knowledge
Obergruppe *f* supergroup
Oberklasse *f* superclass
Objekt *n* object, pattern, frame; entity *{member of a class of objects with the same properties, e.g. a human being}*
Objektbaum *m* context tree, object tree
Objektbeschreibung *f* object description
Objektbeziehung *f* object relation
Objekterkennung *f* object (pattern) recognition

Objekterzeugung *f* object generation
Objektklasse *f* pattern class
objektklassifizierendes System *n* pattern-classifying system
Objektklassifizierung *f* pattern classification
Objektmenge *f* set of patterns
Objektmodell *n* object model
objektorientierte Darstellung *f* declarative representation, object-oriented representation
objektorientiertes Programmieren *n* object-oriented programming, OOP
objektorientiertes System *n* object oriented system, OOS
Objektprogramm *n* target program
Objektraum *m* space of objects
Objektsprache *f* object language, task-oriented language
Objektsystem *n* object system
ODER *n* OR
  ODER-AUCH *n* inclusive OR
  ODER-Baustein *m* OR module
  ODER-Element *n* OR circuit
  ODER-Funktion *f* OR function
  ODER-Gatter *n* OR circuit, OR gate
  ODER-Gatter *n* **mit zwei Eingängen** two-input OR gate
  ODER-Glied *n* OR circuit
  ODER-Knoten *m* OR node
  ODER-Operator *m* OR operator
  ODER-Verknüpfung *f* OR connective
offene gerichtete Pfeilfolge *f* open oriented walk
  offene Kantenfolge *f* open walk
  offene Struktur *f* open structure
  offenes Bedienungssystem open queueing system
  offenes System *n* open system
offensichtlich plausible
OFT OFTEN *{fuzzy quantor }*
ökonomisches Makromodell *n* economical macromodel
ökonomisches Mikromodell *n* economical micromodel
ökonomisches Modell *n* economical model
ökonomisches Spiel *n* business game
Operation *f* operation
Operationsforschung *f* operations research, OR
Operator *m* operator
Opposition *f* contradiction
optimale disjunktive Normalform *f* optimal disjunctive normal (standard) form
optimale Filterung *f* optimal filtering
optimale Kenngröße *f* optimal parameter
optimale Kompromißmenge *f* otimum compromise set, pareto set
optimale Lösung *f* optimum solution
optimale Schaltfunktion *f* optimum switching function
optimale Schätzung *f* optimum estimation
optimale Strategie *f* optimal strategy, optimum strategy
optimale Trajektorie *f* optimal trajectory
optimaler Parameter *m* optimal parameter
optimaler Plan *m* optimum design
optimaler Verlauf *m* optimum behaviour

Optimalfilter 192

optimaler Zustand *m* optimum condition
optimales Filter *n* optimal filter
optimales System *n* optimal system
optimales Verhalten *n* optimum behaviour
Optimalfilter *n* optimal filter
Optimalfilterproblem *n* optimal filtering problem
Optimalität *f* optimality
Optimalitätsbedingung *f* condition for optimality, optimality condition
Optimalitätskriterium *n* criterion of optimality, optimality criterion
Optimalitätskriterium *n* **von Savage** optimality criterion of Savage
Optimalitätsmodell *n* optimality model
Optimalitätsprinzip *n* optimality principle
Optimalitätsprinzip *n* **nach von Neumann** maximin principle
Optimalitätsprinzip *n* **nach von Hurwicz** maximax principle
Optimalprinzip *n* optimum principle
Optimalstrategie *f* optimal (optimum) strategy
Optimalwert *m* optimal value, optimum
Optimalwertschätzung *f* optimum estimation
Optimalwertsystem *n* optimizing system
optimierbar optimizable
Optimierbarkeit *f* optimizability
optimieren optimize/to
optimierender Computer *m* optimizing computer
optimierendes System *n* optimizing system

Optimierregelung *f* adaptive control with optimization *{numerical control}*
Optimierung *f* optimization
Optimierung *f* **auf der Grundlage von Modellen** model-based optimization
Optimierung *f* **auf oberer Ebene** higher-level optimization
Optimierung *f* **auf unterer Ebene** lower-level optimization
Optimierung *f* **bei vorgegebenen Randbedingungen** constrained optimization
Optimierung *f* **der Redundanz** redundancy optimization
Optimierung *f* **der Reihenfolge** sequence optimization
Optimierung *f* **des Elementarsystems** elementary optimization
Optimierung *f* **des Gewinns** gain optimization
Optimierung *f* **durch Dekomposition** decomposition optimization
Optimierung *f* **nach maximalem Gewinn** maximum gain optimization
Optimierung *f* **nach mehreren Zielkriterien** optimization subject to several criteria
Optimierung *f* **nach minimalem Verlust** minimum loss optimization
Optimierung *f* **ohne Modell** direct optimization
Optimierung *f* **von Teilsystemen** subsystem optimization
Optimierung *f* **ohne Randbedingungen** constraint-free optimization
Optimierungsalgorithmus *m* optimization algorithm

Optimierungsaufgabe *f* optimization problem
Optimierungskriterium *n* optimization criterion
Optimierungsproblem *n* optimization problem, optimizing problem
Optimierungsproblem *n* **mit Randbedingungen** constrained optimization problem
Optimierungsproblem *n* **ohne Schranken** unbounded optimization problem
Optimierungsstrategie *f* optimization strategy
Optimum *n* optimal value, optimum
ordentlich right
ordnen order/to, sequence/to
Ordnung *f* order; degree
  Ordnung *f* **eines Baums** degree of a tree
  Ordnung *f* **eines Knotens** degree of a node
Ordnungsbeziehung *f* order[ing] relation
Ordnungsisomorphie *f* isomorphism of order
Ordnungsrelation *f* order relation
Ordnungstendenz *f* ordering bias
Organisation *f* organization; structure *{e.g of a system}*
Organisiertheit *f* state of organization
orthogonal orthogonal
  orthogonale Suche *f* orthogonal search
  orthogonale zentrale Planung *f* orthogonal central design *{experimental design}*
Orthogonalität *f* orthogonality
örtlich local
ortsfest stationary

Ortungswahrscheinlichkeit *f* position-finding probability *{search theory}*
Outputneuron *n* output neuron *{sends information out to the human or to something else}*
Outstar *m* outstar *{a neural structure which samples signals and is used in conditioning}*

P

Paar *n* pair
paarer Graph *m* paired graph
Packungsproblem *n* [bin] packing problem *{graph theory}*
parabolische Optimierung *f* parabolic optimization
paradigmatische Relation *f* paradigmatic relation
parallel parallel, concurrent
  parallel arbeitender Automat (Computer) *m* parallel automaton, fifth-generation computer
  parallele Kanten *fpl* parallel edges
Parallel-Serien-Umsetzer *m* serializer
Parallelbetrieb *m* interleave mode
Parallelcomputer *m* connecting machine, parallel computer (machine)
parallelgeschaltet connected in parallel
Parallelisierung *f* **eines Algorithmus** parallelization of an algorithm
Parallelisierungsstrategie *f* parallelizing strategy
Parallelismus *m* parallelism
Parallelkonflikt *m* parallel-work conflict
Parallelnetz *n* parallel network

Parallelprogrammierung *f* concurrent programming
Parallelprozessor *m* concurrent processor
Parallelprozessorsystem *n* parallel processor system
Parallelschaltung *f* parallel network
Parallelsystem *n* parallel system
Parallelverarbeitung *f* parallel processing
Parallelverarbeitungssystem *n* parallel processing system
Parallelverknüpfung *f* parallel interconnection
Parameter *m* parameter
parameterabhängig parameter-dependent
Parameteradaption *f* parameter adapt[at]ion
parameteradaptives System *n* parameter-adaptive system
parameteradaptiv parameter-adaptive
Parameteränderung *f* parameter fluctuation
Parameteranpassung *f* parameter adapt[at]ion
parameterfrei parameter-free
parameterfreier Test *m* non-parametric test
Parameterklasse *f* parameter class
Parameterraum *m* parameter space
Parameterschwankung *f* parameter fluctuation
parameterunabhängig parameter-independent
parameterunempfindlich parameter-insensitive
Parametervariation *f* parameter variation
Parametervektor *m* parameter vector
Parameterveränderung *f* parameter variation
parametrisch parametric
parametrische Sprachcodierung *f* parametric coding of speech
Pareto-Menge *f* optimum compromise set, Pareto set
Pareto-Optimum *n* Pareto optimum
Parser *m* parser {*computer program for linguistic text analysis*}
Parsing *n* parsing {*linguistic text analysis*}
Partie *f* game {*e.g. chess*}
partiell determinierter Automat *m* partially determined automaton
Partiesituation *f* game situation
partikulärer Quantifikator *m* existential quantifier
Partikularisierung *f* existential quantification
passives Signal *n* at-rest signal
Patient *m* patient
patientenorientierte Daten *npl* patient-oriented data
Patientenüberwachung *f* patient monitoring
Patientenüberwachungssystem *n* patient monitoring system
Pausenzeit *f* rest time {*queueing system*}
Pegel *m* level
Pegelwechsel *m* level change {*logic level*}
Peirce-Funktion *f* Peirce function
Peirce-Pfeil *m* Peirce arrow
Penalty-shifting-Methode *f* penalty function shifting method
Penaltyabschätzung *f* penalty estimation

Penaltyfunktion *f* penalty function
Penaltyfunktionsmethode *f* penalty function method
Penaltyprinzip *n* penalty principle
Performanzsystem *n* performance system
periodisch periodic[al], cyclic
Perzeptor *m* perceptor
Perzeptron *n* perceptron *{a threshold-controlled two-state sensory model created by Rosenblatt}*
Petri-Netz *n* Petri net
Pfad *m* path *{track through a state graph}*
Pfeil *m* arrow, oriented edge
Pfeildiagramm *n* arrow diagram *{graph}*
pfeilorientierter Netzplan *m* arrow-oriented graph
Pflege *f* updating *{file}*
pflegen update/to *{file}*
Phänomen *n* phenomenon
Phonem *n* phoneme
Phonetik *f* phonetics
phonetische Ausgabe *f* acoustic output
phonetische Eingabe *f* acoustic input
Phrase *f* phrase
Phrasenstrukturgrammatik *f* phrase structure grammar
Phrasenstrukturregel *f* phrase structure rule
physikalische Modellierung *f* physical model[l]ing
physikalisches Modell *n* physical model
physikalisches Simulationssystem *n* physical simulation system
Pipeline *f* pipeline
Pipeline-Prinzip *n* pipeline principle
Pipeline-Rechner *m* pipeline computer
Pipeline-System *n* pipeline system
Pipeline-Verarbeitung *f* pipelining
Pivot-Spalte *f* pivot column
Pivot-Spaltensuche *f* pivot column search
Pivot-Suche *f* pivot search
Plan *m* design, schedule, schema
planarer Graph *m* planar graph
planen plant/to, design/to, schedule/to
Planspiel *n* experimental (operational, plan) game
Planung *f* planning, design
Plastizität *f* plasticity *{ability of a group of neurons to adapt their function to a different function over time and to take over the functions of a damaged portion of a neuronal network}*
plausibel plausible
Plausibilitätskontrolle *f* reasonableness check
plausibler Schluß *m* plausible inference
Playback-Roboter *m* [record-]playback robot
Plex *m* plex *{a tree-like data structure}*
Plex-Datenbank *f* plex data base
Plex-Struktur *f* plex structure
Poisson-Verkehr *m* Poisson traffic
Poissonscher Prozeß *m* Poisson process
Poissonscher Strom *m* Poisson stream
Polyautomat *m* polyautomaton
polymorph polymorphic

Polyoptimierung *f* polyoptimization *{optimization with contradictory target functions}*
Polyoptimum *n* polyoptimum
Positionsspiel *n* position game
positive Antwort *f* acknowledgement
Postsektion *f* postsection *{logic}*
Potentialproblem *n* potential problem *{network}*
Prädikat *n* predicate *{logic}*
Prädikatenkalkül *m* predicate calculus
Prädikatenkalkül *m* **erster Stufe** narrow predicate calculus
Prädikatenlogik *f* predicate logic, quantification theory
Prädikatenlogik *f* **erster Ordnung** first-order predicate logic
Prädikatenlogik *f* **höherer Stufen** higher-order logic, higher predicate calculus
prädikatenlogisch allgemeingültig universally valid
prädikatenlogische Formel *f* formula for the predicate calculus
prädikatenlogische Identität *f* predicate-logical identity
prädikatenlogische Sprache *f* language of the predicate calculus, predicate-logical language
prädikatenlogischer Funktor (Junktor) *m* predicate-logical functor
Prädikatenumformer *m* predicate transformer
Prädikatenvariable *f* predicate variable
Prädikativität *f* predicativity
Prädiktion *f* prediction
Prädiktionsalgorithmus *m* prediction algorithm
Prädiktionsfilter *n* predicting filter
Prädiktionsmethode *f* predictive method
prädiktive Codierung *f* predictive coding
Prädiktor *m* predictor
prädizieren predict/to
Präfixnotation *f* prefix notation *{in LISP used list representation}*
präkompakter Raum *m* precompact space
präkompaktes Spiel *n* precompact game
Prämisse *f* premise
Prämissenfolge *f* sequence of premises
pränexe Normalform *f* prenex normal form
pränexer Ausdruck *m* prenex form
Präordnung *f* preordering
Präsektion *f* presection
Präsentationsgraphik *f* business graphics
Präsenz *f* presence
präskriptive Logik *f* prescriptive logic
prävollständige Funktionenklasse *f* precomplete class of functions
Preismethode *f* objective coordination
Primal-Dual-Algorithmus *m* primal dual algorithm, stepping stone method *{linear optimization}*
Primalproblem *n* primal problem *{simplex method}*
primär primary
Primärbeschreibung *f* primary description

primäre Objektbeschreibung *f*
primary description
Primärsprachenübersetzung *f*
source-language translation
Primimplikation *f* prime implication
Primitiv *n* primitive *{semantics}*
primitive Rekursion *f* primitive recursion
primitiv-rekursive Funktion *f*
primitive recursive function
primitiv-rekursive Relation *f*
primitive recursive relation
primitiv-rekursives Prädikat *n*
primitive recursive predicate
Primkonjunktion *f* prime conjunction
Prinzip *n* **der kleinsten Quadrate**
principle of least squares
Prinzip *n* **der maßstabgerechten Verkleinerung** scaling principle
Prinzip *n* **der Umkehrbarkeit**
reciprocity principle
Prinzip *n* **der Zielrealisierbarkeit**
principle of realizability of result
Prinzip *n* **der Zweiwertigkeit**
two-value principle
Prinzip *n* **des Arguments** argument principle
Prinzip *n* **des indirekten Schließens** principle of indirect conclusion
Prinzip *n* **des induktiven Schließens** principle of inductive conclusion
Prinzip *n* **des inneren Modells**
internal-model principle
Prinzip *n* **des kleinsten Bedauerns**
optimality criterion of Savage
Prinzip *n* **vom ausgeschlossenen Dritten** principle of excluded middle (third)

Prinzip *n* **vom ausgeschlossenen Widerspruch** principle of contradiction
Prinzipien *npl* **der Systemanalyse**
principles of systems analysis
Priorität *f* priority
Prioritätscodierer *m* priority coder
Prioritätsebene *f* level of priority, priority level
Prioritätsentscheidung *f* priority arbitration, priority decision
Prioritätsentscheidungslogik *f*
priority arbitration logic
Prioritätsfolge *m* priority sequence (ranking), order of precedence
Prioritätsgrad *m* priority level
Prioritätskreis *m* priority circuit
Prioritätslogik *f* priority logic
Pro-System *n* prospective system, pro system
probabilistisch probabilistic, stochastic
probabilistische Logik *f* probabilistic logic
probabilistisches Modell *n* probabilistic model
probabilistisches System *n* probabilistic system
probierender Automat *m* learning-by-success automaton, trying automaton
Problem *n* problem
Problem *n* **der drei Häuser und drei Brunnen** nine-footway problem, three-houses-and-three-wells problem
Problem *n* **der minimalen Futterkosten** diet problem
Problem *n* **der neun Fußwege**
nine-footway problem, three-houses-and-three-wells problem

Problem *n* **der Verteilung von Lieferungen** problem of the distribution of deliveries
Problem *n* **der zänkischen Nachbarn** nine-footway problem, three-houses-and-three-wells problem
Problem *n* **des billigsten Telefonnetzes** problem of the cheapest telephone network
Problem *n* **des Handelsreisenden** problem of random walk, travelling-salesman problem
Problem *n* **des kürzesten Weges** problem of shortest way
Problem *n* **mit beweglichen Enden** problem with mobile ends
Problem *n* **mit festen Zeitpunkten** problem with fixed instants of time
Problem *n* **mit unvollständiger Information** incomplete-information problem
problemabhängig problem-dependent
Problemanalyse *f* problem analysis
Problemanalytiker *m* problem analyst
Problembereich *m* domain, problem space
problembeschreibende Sprache *f* problem-describing language
Problembeschreibung *f* problem description
Problembestimmung *f* problem determination
problembezogen problem-oriented
problemcharakteristisch problem-characteristic
Problemdefinition *f* problem definition
Problemdefinitionssprache *f* problem definition language

Problemdiagnose *f* problem diagnosis
Problemeingabe *f* problem input
Problemgröße *f* problem variable
Problemklasse *f* problem class
Problemkorrektur *f* problem correction
Problemlösen *n* problem solving
Problemlöser *m* problem solver
Problemlösung *f* problem approach (solution)
Problemlösungsmethode *f* problem solving method
problemnah *s.* problemorientiert
problemorientiert problem-oriented, tailored
problemorientierte (problemnahe) Notation *f* problem-oriented notation
problemorientierte Software *f* problem-oriented software
problemorientierte Sprache *f* problem-oriented language *(programming language)*
problemorientierter Instruktionsvorrat (Befehlvorrat) *m* problem-oriented instruction set
Problemraum *m* problem (search) space
Problemreduktion *f* problem reduction
Problemsituation *f* problem situation
Problemstatus *m* problem state
Problemstellung *f* problem formulation
Problemumformulierung *f* problem reformulation
problemunabhängig problem-independent
Problemvariable *f* problem variable

Problemzeit *f* problem time *{time interval in the simulated system}*
Problemzustand *m* problem state
Produktion *f* production *{IF-THAN rule in expert systems}*
Produktionsregel *f* implication, production rule, IF-THEN rule
Produktionssystem *n* production system, rule-based system
Produktmenge *f* product set
Prognose *f* prognosis, prediction *{prospective system}*
Prognosemethode *f* prediction (prognostic) method
Prognosesystem *n* prognostic method
Prognosewert *m* predictive value
Programm *n* program
Programmabschnitt *m* task
programmgesteuert sequence-controlled
programmierbar programmable
programmierbare Array-Logik *f* programmable array logic
programmierbare Taste *f* program-definable key
programmierbares logisches Entwicklungssystem *n* programmable logical development system
programmierbares logisches Feld *n* programmable logical array
Programmiereingabe *f* **durch Lehren** teach-in *{e.g. industrial robot}*
Programmieren *n* **in Maschinencode** absolute programming *{absolute adresses}*
programmierendes Programm *n* programming program
Programmierhilfe *f* programming tool

Programmiermodul *m* programming module
Programmiersprache *f* programming language
programmiert programmed
programmierte Logik *f* programmed logic
programmiertes Lehrbuch *n* programmed textbook
programmiertes Logikfeld *n* programmed logic array
Programmierumgebung *f* programming environment
Programmierung *f* programming
Programmierung *f* **künstlicher Intelligenz** artificial intelligence programming
Programmiervariable *f* programming variable
Programmierwerkzeug *n* programming tool
Programmodell *n* programming model
progressive Reduktion *f* progressive reduction *{logic}*
Projektierungssystem *n* project design system
Projektion *f* **[in eine andere Dimension]** projection *{mapping an image to another dimension}*
PROLOG PROLOG *{a preferably in expert systems used programming language}*
Proportion *f* proportion
Proportionalität *f* proportionality
Proportionalitätsbereich *m* proportionality range
prospektiver Automat *m* prospective automaton
prospektives System *n* prospective system

Prospektor *m* prospector
Protokoll *n* record
Prototyp *m* prototype
Proximum *n* proximum
Prozedur *f* procedure *{a subprogram}*
prozedural procedural
prozedurale Sprache *f* procedural language
prozedurale Wissensdarstellung *f* procedural knowledge representation
Prozeß *m* process, task
prozeßabhängig process-dependent
Prozeßanalyse *f* process analysis
Prozeßanalyse *f* mittels Netzplantechnik activity analysis
Prozeßanalysenmeßeinrichtung *f* process analyzer
Prozeßbeschreibung *f* process description
Prozeßereignis *n* process event
Prozeßerkennung *f* process recognition
Prozeßinformation *f* process information
Prozeßkomplexität *f* process complexity
Prozeßmodell *n* process model
Prozeßmodellparameter *m* process model parameter
Prozeßparameter *m* process parameter
Prozeßparameterschätzung *f* process parameter estimation
Prozeßsicherung *f* process safeguarding
Prozeßsimulierung *f* process simulation
Prozeßstabilisierung *f* process stabilization
Prozeßsynthese *f* process synthesis
prozeßunabhängig process-independent
Prozeßzustand *m* process state
Prozeßzustandsbeschreibung *f* process state description
prüfen try/to
Prüfmethode *f* method of test
Prüfprogramm *n* für Rechtschreibung spelling checker *{text processing}*
pseudoadaptive Resonanz *f* pseudoadaptive resonance *{occurs when the two layers of a bidirectional associative memory achieve equilibrium}*
Pseudoaddition *f* pseudoaddition *{Boolean algebra}*
Pseudocode *m* abstract code
pseudokonvexe Optimierung *f* pseudoconvex optimization
Pseudomal *n* pseudotimes *{Boolean algebra}*
Pseudoreduktion *f* pseudoreduction
Pseudovariable *f* pseudovariable
Pseudoveränderliche *f* pseudovariable
psychologisches Modell *n* psychological model
Punkt-Strecken-Steuerung *f* straight endpoint-to-point control
Punktgitternetzwerk *n* cutpoint network *{cellular circuit}*
Punktmenge point set
Punktschweißroboter *m* spot welding robot
Pyramidenzelle *f* pyramidal cell *{majority of nerve cells in the cerebral cortex}*

Q

quadratische Optimierung *f* parabolic optimization
Quantifikation *f* quantification
Quantifikator *m* quantifier
quantifizieren quantify/to
quantifizierte Variable *f* quantified variable
quantisiertes System *n* quantized system
Quantisierung *f* truncation
Quantisierungsfehler *m* truncation error
Quantor *m* quantifier, quantor
quasianaloges Modell *n* quasi-analog[ue] model
quasilinearer Automat *m* quasi-linear automaton
quasioptimal quasi-optimum
Quasiordnung *f* preordering
quasiperiodischer Betriebszustand *m* almost periodic behaviour
Quasistabilität *f* quasi-stability
quasistationär quasi-stationary
Quellbibliothek *f* source library
Quelle *f* source
  Quelle *f* **des Turniers** tournament source
  Quelle *f* **ohne Gedächtnis** zero-memory source
Quellenverkehr *m* source traffic
Quellknoten *m* source node
Quellsatz *m* source set
Quellsprache *f* source language
Querschnitt *n* cross section
Querverbindung *f* interconnection *{between neurons}*
quittieren acknowledge/to, handshake/to
Quittung *f* acknowledgement

R

Rahmen *m* [data] frame
Rahmenelement *n* frame element
Randbedingung *f* boundary condition, constraint
Randdichtefunktion *f* boundary density function
randomisieren randomize/to
Randomisierung *f* randomization
Randwertaufgabe *f* boundary value problem
Randwertbedingung *f* boundary condition
Randwertproblem *n* boundary value problem
Rang *m* rank
Rangfolge *f* ranking
Rangordnung *f* priority queue
Raum *m* **der gemischten Strategien** mixed-strategy space
räumlich spatial
räumliches Muster *n* spatial pattern *{at a certain time parallel values of signals}*
Raumstruktur *f* space structure
Rauschen *n* noise *{irrelevant or imprecise data present in input patterns of neural networks}*
Rauschsättigung *f* noise saturation
Rauschunterdrückung *f* noise suppression
Reaktionsmodell *n* model of reaction
realer Automat *m* physical machine
reales Problem *n* real-world problem
Realfall *m* real situation
Rechenanlage *f* computer
Rechenaufwand *m* amount of computation

Rechenautomat m computer, computing automaton (machine)
Rechenautomat m mit beliebiger Befehlsfolge arbitrary-sequence computer
Rechenelement n arithmetic element
Rechenfunktion f arithmetic function
Rechengenauigkeit f accuracy of computation
Rechengeschwindigkeit f arithmetic speed, computing speed
Rechenglied n arithmetic element
Rechenhilfe f computational tool
Rechenlogik f computing logic
Rechenoperation f arithmetic (computing) operation
Rechenoperator m arithmetic operator
Rechenprozeß m task
Rechnercode m absolute code
Rechtschreibprüfprogramm n spelling checker
rechtwinklig orthogonal
Recogniser m recognizer
Record m record *{part of a file}*
Reduktionsschluß m reductive conclusion
reduktiver Schluß m reductive conclusion
redundantes System n redundant system
Redundanz f redundancy *{a deliberate duplication of information, devices or other}*
Redundanzoptimierung f redundancy optimization
reduzierte Grammatik f reduced grammar
reduzierter Automat m reduced automaton

Reduzierung f des [restlichen] Unterschiedes difference reduction *{problem-solving method}*
reell real
referieren abstract/to
reflexiv reflexive
reflexive Relation f reflexive relation
reflexives Spiel n reflexive game
Reflexivität f reflexivity
Regel f rule
regelbasiertes System n production system, rule-based system
Regelbasis f rule base
Regelerzeugung f rule generation
Regelextraktion f rule extraction *{producing the desired output by generalization}*
Regelinterpreter m rule interpreter
Regellogik f mathematical logic
regellos random, stochastic; *{s.a. zufällig}*
 regellos anordnen randomize/to
 regellos auswählen randomize/to
regeln control/to
regelorientiert rule-oriented
Regelschwingung f cycling
Regelung f control
Regelungsstrategie f control strategy
Regelungstheorie f control theory
Regelungsziel n control objective
Regelzweck m control objective
Registerumlauf m cyclic shift
Regression f regression *{statistics}*
regressive Reduktion f regressive reduction
Regularitätsaxiom n axiom of regularity
regulieren control/to
Regulierung f adjustment

Reigenmodell *n* round-robin model *{queueing theory}*
Reihe *f* series, array
Reihe *f* **mit abwechselndem Vorzeichen** alternating series
Reihenfolge *f* order, sequence
Reihenfolge *f* **der Ankunft** arrival sequence
Reihenfolgeoptimierung *f* sequence optimization
Reihenfolgeproblem *n* sequence problem
Reihenmatch *n* matching
Reihenstruktur *f* series structure
reihenweise serial
Reiz *m* cue, stimulus *{any energy input that tends to affect behaviour}*
Reizäquivalenz *f* stimulus equivalence *{causes the ability to determine that an object seen in different orientations or positions is still the same object}*
Reizaufnahme *f* sensing
Reizreaktionsmodell *n* stimulus response model
rekonfigurierbar reconfigurable
rekonstruierbar reconstructable
Rekurrenz *f* recurrence
Rekursion *f* recursion
Rekursionsbeweis *m* proof by recursion
Rekursionsmethode *f* recursive method
rekursiv recursive
rekursiv abzählbare Menge *f* recursively enumerable set, partially decidable set
rekursiv abzählbare Relation *f* recursively enumerable relation
rekursiv abzählbares Prädikat *n* recursively enumerable predicate

rekursive Funktion *f* recursive function
rekursive Operation *f* recursive operation
rekursive Prozedur *f* recursive procedure
rekursiver Prozeß *m* recursive process
rekursives Prädikat *n* recursive predicate
rekursives Spiel *n* recursive game
Relation *f* relation[ship]
relationale Struktur *f* relational structure *{an abstract structure to associate pieces of information}*
relative Priorität *f* head-of-the-line priority *{queueing theory}*
relatives Extremum *n* local extremum
relatives Minimum *n* local minimum
relevante Rückwärtsverkettung *f* relevant backtracking *{to the most relevant choice point but not to the most recent one}*; nonchronological backtracking, dependency-directed backtracking
relevantes Rückwärtsverfolgen *n* (Zurückverfolgen) relevant backtracking
relokatierbar relocatable
repetieren playback/to *{tape}*
Replikation *f* replication
Repräsentationsproblem *n* representation problem *{graph theory}*
reproduzierbar reproducible
Reproduzierbarkeit *f* reproducibility
Reservesystem *n* back-up system
reservieren acquire/to *{e.g. memory}*

Resonanzzustand m resonant state *{resonant activity beween incoming information and feedback expectancies}*
Ressource *f* resource
Ressourcenverteilung *f* resource allocation
Ressourcenverteilungsanalyse *f* resource-distribution analysis
Restverkehr m loss traffic
retrospektive Suche *f* retrospective search
   retrospektiver Automat m retrospective automaton
   retrospektives System n retrospective system
Retrosystem n retrospective system, retro system
Rezeptor m receptor
Rezeptparadigma n prescriptive paradigm
reziprok inverse
rheolineares System n rheolinear system
richtig proper, right
Richtigbefund m acknowledgement
Richtigkeit *f* **des Ergebnisses** accuracy of the result
richtungsunabhängig non-directional
Richtwert m approximation value
Ringverschiebung *f* cyclic shift
Risikoentscheidung *f* decision with risk
Risikosituation *f* risk case (situation)
Roboter m robot
   Roboter m der dritten Generation generation-three robot
   Roboter m der 1,5ten Generation generation-one-point-five robot
   Roboter m der zweiten Generation generation-two robot
   Roboter m mit beschränkter Bewegungsfolge limited-sequence robot
   Roboter m mit Gelenkarm jointed-arm robot
   Roboter m mit in einer Lernphase bei Bewegungsführung durch den Bediener fest gespeichertem Arbeitsablauf record playback robot
   Roboter m mit Kraftkompensation compliant robot system
   Roboter m mit Kugelgelenken spherically jointed robot
   Roboter m mit redundanten Freiheitsgraden redundant-axis roboter
   Roboter m mit Servosteuerung servo-controlled robot
   Roboter m zum Fügen von Werkstücken joining robot
   Roboter m zum Punktschweißen spot-welding robot
   Roboter m zum Sortieren sizing robot
Robotertechnik *f* robot engineering
Robotik *f* robotics
Rohrleitung *f* pipeline
rollen roll/to *{movement of robot}*
Route *f* route
routen route/to
Rückblick m retrospect
Rückführung *f* feedback *{s.a. Rückkopplung}*
Rückhaltzeit *f* detention time
Rückkehr *f* **[zu einem früheren Punkt eines Suchbereiches]** backtracking
Rückkopplung *f* feedback *{the output signals of a neuron are sent back to its own inputs or to the in-*

puts of neurons in a previous layer or in the same layer}
Rückkopplungsfunktion *f* feedback function
Rückmeldung *f* acknowledgement
Rucksackproblem *n* knapsack problem
Rücktransformation *f* back transformation
Rücktransformierte *f* back transform
Rückwärtsargumentieren *n* backtracking
Rückwärtsausbreitung *f* back propagation *{learning method in which an error signal is fed back altering weights to prevent the same error from happening again}*
Rückwärtslesen *n* backward reading
Rückwärtsoptimierung *f* self-adaptation by search
Rückwärtsverketten *n* backward chaining
Rückwärtsverkettungsmethode *f* top-down approach *{hypothesizing and testing}*
Rückwirkungsfreiheit *f* absence of feedback
ruhend stationary
Ruhepotential *n* resting potential *{voltage difference between inside and outside of a nerve cell when nothing is happening}*
Ruhestellung *f* neutral position
Ruhezustand *m* static state
Rundfahrtproblem *n* problem of travelling salesman, shortest-path (-route) method
Rundumsuche *f* round-robin search
Rundung *f* truncation
Rundungsfehler *m* truncation error

S

S-Ausdruck *m* S expression, symbolic expression
S-Funktion *f* sigmoid function *{transfer function with a proportional range and a high and a low saturation limit}*
sachgebietsorientiertes Programmiersystem *n* subject-oriented programming system
Saldierwerk *n* accumulator
Sammelfahrtproblem *n* travelling-salesman problem
Sammlung *f* accumulation
Sattelpunkt *m* saddle point
Sattelpunkttheorem *n* saddle-point theorem
Sättigung *f* saturation
Sättigungsgrenze *f* saturation limit
Sättigungszustand *m* saturation
Satzlehre *f* syntax
Satz *m* rule, law, principle *{s.a. Prinzip}*, sentence
Satz *m* **der Überlagerung** principle of superposition
Satz *m* **vom ausgeschlossenen Dritten** excluded-third rule, tertium non datur
Satz *m* **vom zureichenden Grund** sufficient-reason rule
Satz *m* **von der bedingten Identität** conditional identity rule
Satz *m* **von der Identität** identity rule
Satz *m* **von der oberen Grenze** axiom of the last upper bound
Satz *m* **von der unbedingten Identität** unconditional identity rule
Satz *m* **von Eigenschaften** set of characteristics

Schaltalgebra *f* switching algebra
schaltalgebraisch Boolean
Schaltelement *n* switching element
Schaltfolge *f* switsching sequence
Schaltfolgetabelle *f* sequence table
Schaltfunktion *f* binary logic function, switching function
Schaltglied *n* gate, switching element
Schaltkreis *m* switching circuit
Schaltkreis *m* **mit mehr als zwei möglichen Zuständen** polyflop
Schaltmatrix *f* switching matrix
Schaltnetz[werk] *n* combinational (switching) circuit
Schaltsystem *n* switching system (circuit)
Schalttabelle *f* sequence (switching) table
Schalttheorie *f* switching theory
Schaltung *f* circuit, network
Schaltungsanalyse *f* network analysis
Schaltungsmodell *n* network model
Schaltvariable *f* switching variable
Schaltvorgang *m* switching event
Schaltwerk *n* sequential network
Schaltzeichen *n* circuit (switching) symbol
scharf stringent {*e.g. stability conditions*}
Schaubild *n* graph
Scheinaktivität *f* fictitious activity
Schema *n* schema
Schicht *f* layer {*group of densely interconnected neurons which share a functional feature*}
Schirmbild *n* image
Schlange *f* chain
schlechtdefiniert bad-defined
Schleifenindex *m* cycle index

Schleifenkriterium *n* cycle criterion
schlichter Graph *m* plain graph
Schließen *n* **nach dem gesunden Menschenverstand** common-sense reasoning
Schluchtensuchverfahren *n* gorge search technique {*optimizing*}
Schluß *m* deduction {*logic*}
Schlußregel *f* rule of inference, deduction rule {*logic*}
Schlußrelation *f* inference relation
Schlüsseltext key phrase, cyphertext, keytext
Schlüsselwort *n* keyword
Schlüsselwortzuordnung *f* keyword assignment
Schlußkette *f* inference chain {*logic*}
Schlußfolgern *n* reasoning {*logic*}
Schlußfolgern *n* **mit dem "gesunden Menschenverstand"** common-sense reasoning
Schlußfolgerung *f* reasoning, conclusion
Schlußfolgerungsstrategie *f* control structure (strategy)
Schlußfolgerungssystem *n* inference engine
Schlußfolgerungsvorgang *m* inference procedure
Schmalspursystem *n* narrow system
schneller Computer *m* high-speed computer
Schnittebene *f* cutting plane
Schnittebenenmethode *f* method of intercepted hyperplanes, cutting-plane method
Schnittstelle *f* interface; programming environment
Schnittstelle *f* **zum Menschen** human interface

schöpferische Computeranwendung *f* creative computing
Schreibschrift *f* script
Schrittextremalsystem *n* step-extremal system
Schritt *m* step
Schrittoptimierungsmethode *f* hill-climbing method
Schrittraster *m* step frame
Schrittvektor *m* step vector {*search step*}
schrittweise Approximation *f* approximation by iteration
Schrittweite *f* step width {*search*}
schrumpfen shrink/to
Schulungsprogramm *n* training software
Schürfer *m* prospector
schützen to proof {*data*}
Schutzmaßnahme *f* proof
Schwächung *f* attenuation
Schwärzung *f* density {*film*}
Schwellenwert *m* threshold [value], cut-off value {*internal activation level in a neuron that triggers the output when the value is reached*}
Schwellenwertelement *n* threshold element
schwellenwertfreie Logik *f* non-threshold logic
Schwellenwertfunktion *f* threshold function {*specific transfer function with an yes-or-no output*}
schwingen cycle/to, oscillate/to
Schwingungsart *f* mode
Schwingungsknoten *m* node
Schwingungstyp *m* mode
Seinszeichen *n* existential quantifier
sekundär secondary
Selbstabgleich *m* self-adjustment
selbständig autonomous

selbstanpassend adaptive, autoadaptive, self-adaptive; {*s.a.* adaptiv}
Selbstanpassung *f* [self-]adaptation
Selbstanpassung *f* durch Suche self-adaptation by search
selbstaufzeichnend self-recording
Selbstausgleich *m* self-regulation
Selbstbeobachtung *f* introspection {*man*}
selbstduale Boolesche Funktion *f* self-dual Boolean function
selbsteingestellte Logik *f* mit höchstem Integrationsgrad self-aligned superintegration logic
selbsteinstellend self-adjusting, [self-]adaptive, self-regulating {*the network is able to adapt to changing inputs by modifying its internal processes*}
selbsteinstellender Regler *m* adaptive controller
selbsteinstellendes System *n* goal-searching system, adaptive system
Selbsteinstellung *f* adaptation, self-adjustment, self-adjusting {*the network is able to adapt to changing inputs by modifying its internal processes*}
Selbsterläuterung *f* eines laufenden Programms mid-run explanation
selbstheilendes System *n* self-healing system
Selbstindizierung *f* automatic indexing
selbstlernende Mustererkennung *f* self-learning pattern recognition
selbstoptimierend self-adaptive, self-optimizing
selbstoptimierendes System *n* self-optimizing system

Selbstoptimierung *f* self-optimization
Selbstorganisation *f* self-organization *{self-training of neural networks}*
selbstorganisierender Prozeß *m* self-organizing process
selbstorganisierendes System *n* self-organizing system
selbstprogrammierendes System *n* self-programming system
selbstregistrierend self-recording
Selbstregulation *f* self-regulation
Selbstregulierung *f* self-regulation
Selbstreparatur *f* self-repair
selbstreparierendes System *n* self-healing system, self-repairing system
Selbstreproduktion *f* self-reproduction
selbstreproduzierender Automat *m* self-reproducing automaton
selbstsichernd self-locking
selbstsperrend self-locking
selbststrukturierend self-structurizing
Selbsttest *m* self test
SELTEN SELDOM *{fuzzy quantor}*
Semantik *f* semantics
Semantikfehler *m* semantic error
semantisch semantic[al]
  semantisch äquivalent semantically equivalent
  semantisch vollständig semantically complete
  semantische Analyse *f* semantic analysis
  semantische Äquivalenz *f* semantic equivalence
  semantische Grammatik *f* semantic grammar

  semantische Information *f* semantic information
  semantische Regel *f* semantic rule
  semantische Struktur *f* semantic structure
  semantisches Äquivalenzkriterium *n* semantic equivalence criterion
  semantisches Netz *n* semantic network *{knowledge representation by directed graphs}*
  semantisches Primitiv *n* semantic primitive
semidynamisches System *n* semi-dynamic system
Semi-Markow-Kette *f* semi-Markov chain
Semi-Markow-Prozeß *m* semi-Markov process
Sendeneuron *n* sending neuron
Senke *f* sink
Senke *f* **des Turniers** tournament sink
Sensorneuron *n* sensory neuron, afferent neuron *{supplies signals to the brain from other parts of the body}*
Sensorroboter *m* sensory robot
Sequentialspiel *n* sequential game
sequentiell sequential, serial *{s.a. seriell}*
sequentielle Analyse *f* sequential analysis
sequentielle Computerarchitektur *f* von Neumann architecture
sequentielle Datenstruktur *f* contiguous data structure
sequentielle Entscheidungsfunktion *f* sequential decision function
sequentielle Logik *f* sequential logic

sequentielle Maschine *f* finite (sequential) state machine; finite automaton *{computer with a fixed number of states and rules}*
sequentielle Schaltung *f* sequential circuit, sequential network
sequentielle Suche *f* sequential search
sequentielle Verarbeitung *f* sequential (serial) processing
sequentieller Automat *m* state machine, sequential automaton
sequentielles Spiel *n* sequential game
sequentielles System *n* sequential state machine *{s.a. sequentielle Maschine}*
Sequenz *f* sequence *{logic}*
Sequenzanalyse *f* sequential analysis
seriell serial *{s.a. sequentiell}*
  seriell-parallel serial-parallel
  serielle Arbeitsweise *f* serial mode
  serielle Operation *f* serial operation
  serieller Computer *m* serial computer
  serielles Warteschlangensystem *n* serial queueing system
Serien-Parallel-Serien-Struktur *f* serial-parallel-serial structure
Serien-Parallel-System *n* series-parallel system
Serienarbeitsweise *f* serial mode
Serienbetrieb *m* serial mode, serial (series) operation
Serienstruktur *f* serial structure
Seriensystem *n* series system
Servointegrierer *m* decision integrator, saturating integrator
Shannon *n* Shannon *{in information theory measure of information used similarly to the bit}*
Sheffer-Function *f* Sheffer function
sicheres Ereignis *n* sure event
Sicherheit *f* [confirmative] certainty
Signal *n* signal
Signalanalyse *f* signal analysis
Signalelement *n* signal element
Signalerkennbarkeit *f* intelligibility of signal
Signalleitung *f* connection *{is analogous to a synapse in a biological neuron}*
Signalmenge *f* signal set
Signalmuster *n* signal pattern
Signalpegel *m* level
Signalraum *m* signal space
Signalspannung *f* level
Signalstruktur *f* signal structure
Signalverlauf *m* **nach dem [erwünschten] Endpunkt** after end-point action
Signalvorrat *m* [available] signal supply
Signumfunktion *f* sign function
Simulation *f* simulation *{a representation of a system that imitates the behaviour of the system}*
Simulationsmethode *f* simulation method
Simulationsmodell *n* simulation model
Simulationsparameter *m* simulation parameter
Simulationssprache *f* simulation language
Simulationssystem *n* simulation system
simulieren simulate/to
simuliertes Glätten *n* simulated ironing *{used to remove wrinkles from the E surface}*

Simulierung 210

simuliertes Neuron *n* simulated neuron, siron
simuliertes Vergüten *n* simulated annealing *{used to smooth the E surface and escape the local minima}*
Simulierung *f* simulation
Simulierung *f* **des kontinuierlichen (stetigen) Systems** continuous-system simulation
simultan simultaneous
Simultanbetrieb *m* simultaneous operation
Simultancomputer *m* parallel computer
Simultanität *f* simultaneity
Simultanoperation *f* simultaneous operation
Simultanverarbeitung *f* simultaneous processing
singulär singular
Sinn *m* signification
Sinnbild *n* symbol
Sinnesorgan *n* sensory organ
Sinneswahrnehmung *f* perception
sinnloses Programm *n* demented program
Sinuskurvenmensch *m* sine curve man *{historic computer graphic of a man by sine curves}*
Siron *n* simulated neuron, siron
Situation *f* situation, case
Situationserkennung *f* situation recognition
Situationserkennungsalgorithmus *m* situation recognition algorithm
Skalenanfangswert *m* minimum value
Skalierbarkeit *f* scalability
Skalierungsfaktor *m* gain
Skript *n* script *{e.g. frame-like structure for event sequences}*

Slot *n* slot
Soliduslinie *f* solidus
Sollbahn *f* set path *{numerical control}*
Solltrajektorie *f* set trajectory
Sollwert *m* set[-point] value
Soma *n* soma *{the place of a neuron where incoming signals are added}*
SOPS subject-oriented programming system
Sortieralgorithmus *m* sort algorithm
Sortierbaum *m* sorting tree
sortieren/der Reihenfolge nach sequence/to
Sortiernetzwerk *n* sorting tree
Sortierroboter *m* sizing robot
Spalte *f* column *{matrix}*
spannungsführend active
speicherbehaftetes System *n* system with memory
speicherfrei memory-free, memoryless
Speicherhierarchie *f* memory hierarchy
speicherloser Automat *m* automaton without memory
Speichernetz *n* memory network
Speicherplatte *f* target
Spektrallogik *f* spectrum logic
sperren interlock/to
Speziallogik *f* special-purpose logic
Spezialroboter *m* special[-type] robot
spezielle Funktion *f* special function
Spiegeln *n* mirroring
Spiel *n* game
Spiel *n* **auf dem Einheitsquadrat** game on the unit square
Spiel *n* **auf einem Graphen** game on a graph

Spiel *n* **gegen die Natur** game against nature, statistical game
Spiel *n* **in der Planung und Leitung** business game
Spiel *n* **mit endlichem Baum** finite-tree game
Spiel *n* **mit entgegengesetzten Interessen** contrary-interest game
Spiel *n* **mit nichtentgegengesetzten Interessen** non-contrary interest game
Spiel *n* **mit Sattelpunkt** saddle-point game
Spiel *n* **mit unendlichem Baum** infinite-tree game
Spiel *n* **mit unvollständiger Information** incomplete-information game
Spiel *n* **ohne Nebenauszahlungen** game without side payment
Spiel *n* **ums Überleben** game of survival
Spiel *n* **zur Wahl eines Zeitpunktes** game of timing
Spiel *n* **über dem Einheitsquadrat** unity-square game
spielender Automat *m* playing automaton
Spieler *m* player *{theory of games}*
Spielmatrix *f* game matrix
Spielmodell *n* game model
Spielsimulierung *f* game simulation, gaming
Spielsituation *f* game situation
Spielstrategie *f* strategy of game
spieltheoretisches Modell *n* game-theoretic model
Spieltheorie *f* game theory, theory
Spielverhalten *n* game behaviour
Spielwert *m* game value
Spielzahl *f* cycle criterion

Sprachanalysator *m* parser, speech analyzer
Sprachanalyse *f* language analysis; speech analysis
Spracharchitektur *f* language architecture
Sprachausgabe *f* acoustic output, audio response unit, speech output
Sprachbaustein *m* speech module, speech module
Sprache *f* speech; language *{abstract}*
Sprache *f* **höherer Ordnung** high-order language
Spracheingabe *f* acoustic input, speech input
Spracherkennung *f* speech recognition
Spracherkennungssystem *n* speech recognition system
Sprachinterpreter *m* language interpreter
Sprachkommunikation *f* speech communication
Sprachkompression *f* speech compression
Sprachmodell *n* language model
Sprachmodul *m* speech module
Sprachredundanz *f* language redundancy
Sprachsignalsynthese *f* synthesis of speech signals
Sprachsimulation *f* speech simulation
Sprachsynthese *f* speech synthesis
Sprachübersetzer *m* language translator, assembler
sprachverarbeitender Computer *m* language-processing computer
Sprachverständnis *n* speech understanding

Sprachverstehen n speech recognition *{by computer}*
Sprachverstümmler m speech scrambler
sprecherabhängige Spracherkennung f speaker-dependent speech recognition
sprecherabhängiges Erkennungssystem n speaker-dependent recognition system
sprecherunabhängige Spracherkennung f speaker-independent speech recognition
Sprühkopf m spray head *{paint-spraying robot}*
Sprung m step *{signal}*; level change *{voltage}*
stabil stable
  stabiler Bereich m region of stability
  stabiler Knotenpunkt m stable nodal point
  stabiler Zustand m stable state
  stabiles System n stable system
Stabilität f stability
  Stabilität f eines Systems system stability
  Stabilität f im Großen stability in the large
  Stabilität f im Kleinen stability in the small
Stabilitätsbereich m region of stability
Stabilitätsgrenze f boundary of stability, critical stability, limit of stability
Stabilitätskriterium n condition for stability, criterion of stability
Stabilitätsproblem n problem of stability
Stand m der Technik state of the art

Standardwert m default value
Standortproblem n site problem
stark zusammenhängender Graph m strongly connected graph
starke Konvergenz f strong convergence
starker Zusammenhang m strong connection *{graph}*
starkes Extremum n strong extreme value
Stärke f strength *{of a rule}*
starr fortlaufend organisierter Computer m consecutive-sequence computer
Startereignis n starting event
Startknopf m activation button
Startmenü n starting menu
stationär stationary, *{s.a. statisch}*
  stationäre Ausgangsgröße f steady-state output
  stationärer Prozeß m steady-state process
  stationärer Wert m steady-state value, conservative value
Stationaritätsbedingung f steady-state condition
statisch static, steady-state
  statische Lösung f steady-state solution
  statische Optimierung f steady-state optimization
  statische Simulierung f steady-state simulation
  statischer Zustand m static state
  statisches Optimierungsproblem n steady-state optimizing problem
  statisches Signal n pattern
statistisch verteilt random
statistische Hypothese f statistical hypothesis

statistisches Modell *n* random model, statistical model
statistisches Schätzungsverfahren *n* statistical method of estimation
statistisches Spiel *n* game against nature, statistical game
Stauung *f* congestion
Steigungsschema *n* gradient scheme *{divided-differences metod}*
Stelle *f* column
Stelleinrichtung *f* actuating unit
Stellelement *n* effector
Stellenwert *m* column
Stellenzahl *f* valence *{logic operation}*
Stellglied *n* actuating element (unit)
Stellorgan *n* effector
Stepping-stone-Methode *f* stepping stone method, primal dual algorithm *{linear optimization}*
stereotype Situation *f* stereotyped (repetitive) situation
Stetigbahnroboter *m* continuous-path robot, path robot
stetig continuous, steady
  stetige Funktion *f* continuous function
  stetige Größe *f* continous variable
  stetiger Entscheidungsprozeß *m* continuous decision process
  stetiger Markow-Prozeß *m* continuous Markovian process
  stetiges Neuron *n* continuous-state neuron
  stetiges System *n* continuous system
Stetigkeit *f* continuity *{mathematics}*
Stetigkeitsaxiom *n* axiom of continuity
steuern control/to
Steuerneuron *n* control neuron, efferent neuron *{sends signals from the brain to other parts of the body}*
Steuerraum *m* control space *{control theory}*
Steuerstrategie *f* control strategy
Steuerstruktur *f* control structure
Steuerung *f* control
Steuerung *f* **nach vorausberechnetem Weg** computed-path control *{robot}*
Steuerungshierarchie *f* control hierarchy
Steuerungsmethode *f* control method *{e.g. forward chaining or backward chaining}*
Steuerungssystem *n* control system
Steuerungssystem *n* **mit Modellvergleich** model-reference control system
Steuerungssystem *n* **mit veränderlicher Struktur** control system with variable structure
Steuerungssystem *n* **mit verteilten Parametern** control system with distributed parameters
Steuerungstheorie *f* control theory
Steuerungsziel *n* control objective
Stichwort *n* keyword
Stimmenerkennung *f* voice recognition
Stimulus *m* stimulus, input, cue, *{any energy input that tends to affect behaviour}*
stochastisch stochastic, probabilistic
stochastische Gewichtsverringerung *f* stochastic weight decay
stochastische Optimierung *f* stochastic optimization
stochastische Quasigradientenmethode *f* stochastic quasi-gradient method

stochastische Simulation *f* stochastic simulation
stochastischer Automat *m* probabilistic automaton, stochastic automaton
stochastisches Spiel *n* stochastic game
stochastisches System *n* stochastic system
stochastisches Verhalten *n* stochastic behaviour
störanfällig accident-sensitive
Störeffekt *m* disturbing effect
störendes Rauschen *n* disturbing noise
Störung *f* trouble
Störungsdiagnose *f* trouble diagnosis
Störungsortungsproblem *n* trouble-location problem
Strafabschätzung *f* penalty estimation
Straffunktion *f* penalty function
Straffunktionsmethode *f* penalty function method
Strafmethode *f* penalty method
Strafprinzip *n* penalty principle
Strafverschiebungsmethode *f* penalty function shifting method
Strategie *f* strategy
Strategie *f* **des Spiels** strategy of game
Strategiecomputer *m* strategic (strategy) computer
Strategiemenge *f* strategy set
Strategieplanung *f* strategy planning
Strategieraum *m* strategy space
Strategieunbestimmtheit *f* indeterminacy of strategy
strategische Äquivalenz *f* strategic equivalence
strategische Entscheidung *f* strategic decision
strategische Situation *f* strategic case, strategic situation
strategisches Modell *n* strategic model
strategisches Spiel *n* strategic game
strategisches Verhalten *n* strategic behaviour
strategisches Wissen *n* strategic knowledge
streckengesteuerter Roboter *m* [continuous-]path robot
strikte Implikation *f* strict implication *{logic}*
Strom *m* current; stream; flow *{graph}*
stromführend active
Stromintensität *f* stream intensity *{arrival process}*
Stromstärke *f* current
Strömung *f* flow, stream
Struktur *f* structure
Strukturanalogie *f* structural analogy
Strukturanalyse *f* structural analysis
Strukturäquivalenz *f* structural equivalence
Strukturattribut *n* structural attribute
Strukturebene *f* structural level
strukturell structural
strukturelle Analogie *f* structural analogy
strukturelle Äquivalenz *f* structural equivalence
strukturelle Automatentheorie *f* structural theory of automata

strukturelle Mehrdeutigkeit *f* structural ambiguity
strukturelle Optimierung *f* structural optimization
strukturelle Redundanz *f* structural redundancy
strukturelle Semantik *f* structural semantics
strukturelle Veränderung *f* structural variation
strukturelle Stabilität *f* structural stability
strukturelles Modell *n* structural model
Strukturerkennung *f* pattern recognition
Strukturerkennungsgerät *n* pattern recognizer
strukturierbar structurable
strukturieren structurate/to
strukturiert structured
  strukturierte Menge *f* structured set
Strukturindex *m* structure index
strukturinstabiles System *n* structurally instable system
Strukturinstabilität *f* structural instability, structure instability
Strukturmatrix *f* structural matrix, structure matrix
Strukturmodell *n* structure model
Strukturmuster *n* pattern
Strukturoptimierung *f* structural optimization, structure optimization
Strukturparameter *m* structure parameter
Strukturproblem *n* structure problem
Strukturredundanz *f* structural redundancy
Strukturregel *f* structure rule

Struktursimulator *m* structure simulator
strukturstabiles System *n* structurally stable system
Strukturstabilität *f* structural stability
Struktursynthese *f* structural synthesis
Strukturtheorie *f* structure theory
Strukturtheorie *f* **von Automaten** structural automata theory
Strukturveränderung *f* structural variation
Stufe *f* step
Stufengewinn *m* step gain *{dynamic optimizing}*
Stufenkalkül *m* level calculus *{predicate logic}*
Stufenprozeß *m* step process *{dynamic optimizing}*
Stützebene *f* plane of support
Stutzen *n* pruning
Stützstelle node
Subgraph *m* subgraph
Subjekt *n* subject
subjektive Sicherheit *f* certainty
Submatrix *f* submatrix
Submodell *n* submodel
suboptimal suboptimum
  suboptimale Lösung *f* suboptimum solution
  suboptimale Strategie *f* suboptimum strategy
Suboptimierung *f* suboptimization, suboptimizing
Substitution *f* substitution
Substitutionsfehler *m* substitution error
Subsystem *n* subsystem
Suchalgorithmus *m* searching algorithm

Suchalgorithmus *m* **von Box-Wilson** Box-Wilson searching algorithm
Suchbaum *m* search tree
Suchbereich *m* searching range
Suche *f* search
   Suche *f* **in die Breite** breadth-first search
   Suche *f* **in die Tiefe** depth-first search
   Suche *f* **mit bedingter begrenzter Semantik** conditional limited semantics search
   Suche *f* **nach dem Gradientenverfahren** gradient search
   Suche *f* **nach Faustregeln** heuristic search
suchen search/to, abstract/to
   suchen/einen Weg route/to
Suchfilter *n* searching filter
Suchfrage *f* query
Suchfrequenz *f* searching frequency
Suchkreis *m* search loop *{backward optimizing}*
Suchlauf *m* search *{e.g. machine tool}*
Suchoperator *m* search operator
Suchprozeß *m* search[ing] process
   Suchprozeß *m* **mit durch Fibonacci-Zahlen bestimmter Anzahl Tests** Fibonacci search process
Suchraum *m* problem space, search space
Suchschleife *f* search cycle *{optimizing}*
Suchschritt *m* search[ing] step
Suchschrittmethode *f* hill-climbing method
Suchstrategie *f* searching strategy
Suchverfahren *n* searching procedure (technique)
Suchverfahren *n* **nach dem Goldenen Schnitt** golden section search procedure
Suchvorgang *m* search
Suchzeit *f* search period
Suchzyklus *m* search[ing] cycle
Summationsfunktion *f* summation function *{an activation function; signals are added and compared against an internal threshold value}*
Summenfeld *n* accumulator
Super-Superzeichen *n* second-order supercharacter
Superpositionsprinzip *n* principle of superposition
Supervisorcomputer *m* coordinator
Superzeichen *n* supercharacter
Superzeichen *n* **erster Ordnung** first-order supercharacter
Superzeichen *n* **zweiter Ordnung** second-order supercharacter
Surjektion *f* surjective function *{set theory}*
Syllogismus *m* syllogism *{deductive logic argument whose conclusion is supported by two premises}*
Syllogistik *f* syllogistics
Symbol *n* symbol, character
Symbol *n* **mit fester Bedeutung** constant
Symbolbibliothek *f* symbol library
Symbolfolge *f* symbol sequence
Symbolik *f* symbology, symbolics
symbolische *f* **Bezeichnung** *f* symbolic name
symbolische Darstellung *f* symbolic representation
symbolische Einheit *f* symbolic unit
symbolische Logik *f* mathematical logic

symbolische Schreibweise *f* symbolic notation
symbolische Übereinstimmung *f* symbolic concordance
symbolischer Name *f* symbolic name
symbolischer Parameter *m* symbolic parameter
symbolischer Prozeß *m* symbolic process
symbolisches Modell symbolic (abstract) model
symbolisieren symbolize/to
Symbolkette *f* symbol string
Symbolliste *f* symbolic listing
Symbollogik *f* mathematical (symbolic) logic
Symbolsprache *f* symbolic language
Symboltabelle *f* symbol dictionary, symbol table
Symbolvariable *f* symbol variable
Symbolverarbeitung *f* symbol processing, symbolic programming (computation)
Symbolvorrat *m* symbol supply
symmetrieren symmetrize/to
Symmetrierung *f* symmetrization
symmetrische Boolesche Funktion *f* symmetric Boolean function
symmetrische Differenz *f* symmetric difference *{set theory}*
symmetrische Gruppe *f* symmetric group
symmetrische Matrix *f* symmetric matrix
symmetrische Relation *f* symmetric relation
Synapse *f* synapse *{area of electrochemical contact between two neurons}*
synchrone Logik *f* synchronous logic

synchroner Automat *m* synchronous automaton
Synentropie *f* synentropy, average transinformation content
Synergetik *f* synergetics
syntagmatische Relation *f* syntagmatic relation
syntaktisch vollständig syntactically complete
syntaktische Analyse *f* syntactical analysis
Syntaxanalysator *m* syntactic analyzer
Syntax *f* syntax
Syntaxfehler *m* syntactic error
syntaxgesteuerter Übersetzer *m* syntactically controlled translator
Syntaxkontrolle *f* syntactic program checking
Syntaxprüfer *m* syntax checker
Synthese *f* synthesis, generation
Synthese *f* **diskreter Automaten** synthesis of discrete automata
Syntheseproblem *n* synthesis problem
Synthesetest *m* synthesis test
synthetische Relation *f* synthetic relation
synthetische Sprache *f* synthetic language
synthetisierbar synthetizable
System *n* system
System *n* **mit automatischer Stabilisierung** automatic stabilization system
System *n* **mit Bezugsmodell** model-reference system
System *n* **mit einem Freiheitsgrad** single-degree-of-freedom system
System *n* **mit internem Modell** internal-model system

System *n* mit konstanten Koeffizienten constant-coefficient system
System *n* mit mehreren Freiheitsgraden system with several degrees of freedom
System *n* mit Selbsteinstellung adaptive control system
System *n* mit verteilten Parametern system with distributed parameters
System *n* mit verteilter Intelligenz distributed-intelligence system, distributed system
System *n* mit verzweigten Parametern hereditary system
System *n* mit vielen Bedienungsapparaten many-server system
System *n* mit zeitlich veränderlichen Parametern rheolinear system
System *n* mit zufällig schwankenden Parametern random parameter system *{at least one parameter}*
System *n* mit zwei [stabilen] Zuständen two-state system
System-Umwelt-Kommunikation *f* system-environment communication
System *n* zum Speichern und Wiederauffinden von Informationen storage-and-information-retrieval system
Systemabfrage *f* system consultation *{expert system}*
systemabhängig system-dependent
Systemabhängigkeit *f* system dependency
Systemanalyse *f* system analysis
systematisch systematic
systematische Ordnungsabweichung *f* ordering bias

Systemausfall *m* system failure
Systembeschreibung *f* system description
Systembestandteil *m* system component
systemdynamisch system-dynamic
Systementwurf *m* system design
Systemgleichung *f* system equation
Systemhierarchie *f* system hierarchy
Systemidentifikation *f* system identification
Systemidentifizierung *f* system identification
Systemklasse *f* system class
Systemkonzept *n* system concept
Systemlösung *f* system approach
Systemmatrix *f* system matrix
Systemmodell *n* system model
Systemmodellierung *f* system model[l]ing
Systemmodifikation *f* modification
Systemoptimierung *f* system optimization
Systemstruktur *f* system structure
Systemsynthese *f* system synthesis
Systemtheorie *f* control theory, system theory
Systemumgebung *f* system environment
Systemverhalten *n* system performance
Systemverklemmung *f* deadlock
Systemverwaltung *f* system management
Systemwirksamkeit *f* system effectiveness
Systemzustand *m* system state
Systemzuverlässigkeit *f* system reliability
systolisches Array *n* systolic array
Szenenanalyse *f* computer vision

T

Tafel *f* blackboard
Tafelmethode *f* blackboard approach *{problem solving}*
Taktdauer *f* timing
Taktgabe *f* timing
Taktik *f* tactics
taktische Planung *f* tactical planning *{short-term}*
Taktzeit *f* timing
Tandembedienungssystem *n* two-phase queueing system
Tastfähigkeit *f* tactile recognition capability
Tastsinn *m* tactile recognition capability, tactile sense *{robot}*
Tätigkeit *f* action
Tatsache *f* fact
tatsächlich actual
tatsächliches Verhalten *n* actual operation, actual performance
Tautologie *f* tautology
tautologisch tautological
Taxonomie *f* taxonomy
Taylor-Entwicklung *f* **Boolescher Funktionen** Taylor series expansion of Boolean functions
Teach-in-Verfahren *n* teach-in method *{robotics}*
Teachware *f* teachware
technische Kybernetik *f* technical cybernetics
Teil *n (m)* part
Teilausdruck *m* partial expression *{logic}*
Teilen *n* sharing
Teilgraph *m* subgraph
Teilhaberbetrieb *m* transaction processing
Teilklasse *f* subclass *{set theory}*

Teilkoalition *f* partial coalition *{theory of games}*
Teilkörper *m* subfield *{mathematics}*
Teilmenge *f* subset, partial set *{set theory}*
Teilplan *m* subplan
Teilproblem *n* subproblem *{secondary problem that must be solved to solve the original problem}*
Teilsystem *n* subsystem
Teilsystemstruktur *f* subsystem structure
Teilverknüpfung *f* partial conjunction *{logic}*
teilweise partial
teilweise geordnete Menge *f* partially ordered set
Teilziel *n* subgoal
Teleoperator *m* teleoperator
Temperatur *f* temperature *{in the context of neural networks an explicitly decreasing function of time}*
Tensor *m* tensor *{vector}*
Term *m* term
Terminal *n* **für Dialogbetrieb** conversational terminal
Terminalknoten *m* leaf node *{last node of a tree}*
terminologische Kontrolle *f* terminological check
ternär ternary
ternäre Logik *f* ternary logic, three-state logic
ternäre Schaltalgebra *f* ternary switching algebra
tertium non datur excluded-third rule
Themenbaum *m* object (context) tree
Themenbereich *m* domain

Theorem *n* theorem
Theoretiker *m* theoretician
theoretisch theoretic[al]
Theorie *f* theory
  Theorie *f* **der endlichen Automaten** finite-automata theory
  Theorie *f* **der logischen Netze** theory of logical nets
  Theorie *f* **der optimalen Filter** theory of the optimum filter
  Theorie *f* **der Schaltnetzwerke** switching theory
  Theorie *f* **der Zufallsprozesse** theory of random processes
  Theorie *f* **lernender Systeme** learning-system theory
Therapie *f* therapy
Tiefe *f* depth
Tiefeninformation *f* depth cue
Tiefensuche *f* deep knowledge, depth-first search
Todesprozeß *m* death process *{random process}*
Tonsystem *n* audio system
Tool *n* tool *{e.g. special software}*
Top-down-Logik *f* top-down logic
Top-down-Methode *f* top-down method
Topologie *f* topology *{mathematics}*
topologische Abbildung *f* homomorphism
topologischer Raum *m* s. abstrakter Raum
Tor *n* gate
Torschaltung *f* gate circuit
Totalausfall *m* total malfunction
totale Black-box *f* total black box
Tourenproblem *n* tour[ing] problem *{linear optimizing}*
Trajektorie *f* trajectory
  Trajektorie *f* **im Zustandsraum** state-space trajectory
Transformationsregel *f* conclusion rule *{logic syntax}*
Transinformation *f* transinformation
Transinformationsgehalt *m* transinformation content
transitiv transitive
  transitive Abhängigkeit *f* transitive dependence
Transitivität *f* transitivity
Transport *m* advance, transport[ation]
Transportnetz *n* transport network *{graph}*
Transportoptimierung *f* transport[ation] optimization
Transportproblem *n* distribution problem, transport[ation] problem
  Transportproblem *n* **mit Kapazitätsbeschränkung** capacity-restricted transportation problem
Treffer *m* hit
Trend *m* trend
Trennung *f* decoupling
Trial-and-error-Methode *f* trial-and-error method
Trieb *m* drive *{a biological important stimulus based on animal needs such as hunger, fear etc.}*
trivalent ternary
Trugschluß *m* fallacy *{logic}*
Tse *n* tse *{chinese character}*
  Tse-Computer *m* tse computer *{pattern recognition}*
Turing-aufzählbare Relation *f* Turing-enumerable relation
Turing-Automat *m* determinated machine, Turing machine
Turing-entscheidbares Prädikat *n* Turing-decidable predicate

Turing-Maschine *f* determinated machine, Turing machine
Turing-System *n* determinated system
Turing-Tafel *f* Turing table
Turnier *n* tournament
Tutorsystem *n* tutor system
typische Menge typical set

U

überaktiv hyperactive *{excessively active}*
überbestimmtes System *n* inconsistent system
überdecken cover/to
Überdeckung *f* mask/to *{a second stimulus is not perceived due to persistent processing a first stimulus}*; overshadowing *{effect of one conditioned response on another}*
Überdeckungsproblem *n* covering problem *{graph theory}*
Überdeckungsraum *m* *{Graph}* covering space *{graph}*
übereinstimmend conformal
Übereinstimmung *f* coincidence, concordance, conformity, correspondence
Überfläche *f* hypersurface
überfliegen browse/to *{text}*
überflüssige Information *f* superfluous information
Überfüllung *f* congestion
Übergangsgraph *m* transition graph
Übergangswahrscheinlichkeit *f* conditional probability
übergeordnet master
übergeordneter Computer *m* higher-level computer, master computer
übergeordnetes System *n* master system
Übergruppe *f* supergroup
überlagern overlay/to
Überlagerungsbaum *m* overlay tree
überlappte Verarbeitung *f* concurrent processing
Überlappung *f* interleave
Überlaufsystem *n* overflow system *{queueing theory}*
Überlaufverkehr *m* overflow traffic *{queueing theory}*
Überlebensfähigkeit *f* survivability *{reliability theory}*
Überlebenswahrscheinlichkeit *f* surviving probability *{reliability theory}*
Überlebenswahrscheinlichkeitsverteilung *f* surviving probability distribution
Übermatrix *f* hypermatrix
Überraschungsgehalt *m* surprise content *{information}*
Überraum *m* hyperspace
Übersetzer *m* interpreter; translator
Übersetzung *f* translation
Übersetzung *f* **natürlicher Sprache** natural language translation
Übersetzungsprogramm *n* translator
Übersetzungssprache *f* translation language
Übersicht *f* schema
überstreichen cover/to
Übertragungsfaktor *m* transfer factor
Übertragungsfunktion *f* transfer function
Übertragungsglied *n* element, link

Übertragungsweg *m* circuit, trunk
überwachtes Lernen *n* supervised learning *{an external influence tells by comparing the network whether or not its output was correct and the weights are adjusted to reduce any difference}*
Umbau *m* modification
Umcodierung *f* conversion
umfassend global, synoptic
Umformung *f* transformation
Umgebung *f* environment *{e.g. programming environment}*
Umgebungseigenschaft *f* feature *{a particular characteristic of the environment, e.g. colour or sound}*
Umgebungsinstrument *n* environment tool
umgekehrt inverse
umkehrbar invertible, reversible
umkehrbar eindeutig one-to-one *{s.a. eineindeutig}*
umkehrbar eindeutiger Homomorphismus *m* one-to-one homomorphism
umkehrbare Funktion *f* invertible function
umkehrbares Modell *n* revertible model
Umkehrbarkeit *f* invertibility
Umkehrbarkeitsprinzip *n* reciprocity principle
umkehren invert/to
Umkehrung *f* inversion *{e.g. logic}*
umkonfigurierbar reconfigurable
Umlauf *m* circuit; turn *{graph}*; cycle
umleiten route/to *{data flow}*
umordnen move/to *{data}*
Umrechnung *f* conversion
umschreiben copy/to

Umsetzung *f* conversion, translation
umspeichern copy/to
Umwandlung *f* conversion, translation
Umweltmodell *n* environmental model
unabhängig autonomous, independent *{system}*
unabhängige Ereignisse *npl* independent events
unabhängiger Wartezustand *m* asynchronous disconnected mode
unäre Operation *f* unary operation
unärer Operator *m* unary operator
unbedingte Anweisung *f* imperative statement
unbedingte Erwartung *f* unconditional expectation
unbedingter Ausdruck *m* unconditional expression
unbedingter Reiz (Stimulus) *m* unconditioned stimulus *{normal stimulus evoking an unconditioned response}*
unbedingtes Verhalten *n* unconditioned response *{normal response to a stimulus, e.g. fear at the sight of fire}*
unbefugter Eingriff *m* tampering
unbegrenzt infinite
unbegrenzter Bereich *m* unbounded domain
unbekannt unknown
Unbekannte *f* unknown quantity
unbekannte Größe *f* unknown quantity
unberechenbar non-computable
unbeschränkter Endzustand *m* unconstrained finite state
unbesetzter Platz *m* unoccupied place *{queueing theory}*

unbestimmt indeterminate; unknown; indifferent
unbestimmte Situation *f* indetermined situation
unbestimmte Stabilität *f* indifferent stability
unbestimmtes Gleichgewicht *n* indifferent stability
Unbestimmtheit *f* indeterminacy, indifference
Unbestimmtheitsmaß *n* measure of indeterminacy, coefficient of indetermination
UND AND
  UND-Glied *n* AND gate
  UND-Knoten *m* AND node
  UND-NICHT-Tor *n* inhibitory gate
  UND/ODER-Graph *m* AND/OR graph *{tree-like structure with AND nodes and OR nodes}*
  UND-Operation *f* conjunction
  UND-Schaltung *f* AND circuit
  UND-Verknüpfung *f* AND operation
uneindeutig fuzzy
Unempfindlichkeitsbereich *m* neutral zone
unendlich infinite
  unendlich-dimensional infinite-dimensional
  unendliche Gruppe *f* infinite group
  unendlicher Automat *m* infinite automaton (machine)
  unendlicher Graph infinite graph
  unendliches Spiel *n* infinite game
Unendlichkeit *f* infinity
Unendlichkeitsaxiom *n* axiom of infinity
Unendlichkeitsprinzip *n* existence-of-infinite-set principle *{set theory}*
unendlichwertige Logik *f* infinite-valued logic
unentscheidbar indecidable, non-decidable, undecidable
Unentscheidbarkeit *f* indecidability
unerlaubter Bereich *m* illegal region *{search}*
unexakt fuzzy, *{s.a. unscharf}*
ungelöster Konflikt *m* unsolved conflict
Ungenauigkeit *f* inexactness
ungerichtet non-directional, random
ungerichtete Kante *f* non-directed edge *{graph}*
ungerichteter Baum *m* non-directed tree *{graph}*
ungerichteter Graph *m* non-directed graph, non-oriented graph
Ungewißheit *f* incertainty, uncertainty
ungewollter Programmabbruch *m* abnormal termination
ungültig invalid
Unifikation *f* unification
Unifizierung *f* unification *{attempt to find substituations for variables that make two atoms identical}*
Universalaussage *f* universal sentence *{logic}*
Universalität *f* universality
Universalroboter *m* general-purpose robot
universelle Funktion *f* universal function
universelle Turing-Maschine *f* universal Turing machine
universeller Algorithmus *m* universal algorithm
universeller Logikbaustein *m* universal logic module
universeller Quantifikator *m* universal quantifier *{logic}*

unkonditionierter Stimulus *m* unconditioned stimulus
unkorrigierbarer Fehler *m* uncorrectable error
unkritischer Punkt *m* non-critical point
unlösbares Problem *n* unsolvable problem
Unmenge *f* second-level set *{set theory}*
unmögliches Ereignis *n* impossible event
Unmöglichkeit *f* impossibility
Unmöglichkeitsbeweis *m* disproving *{attempt to prove the impossibility of a hypothesized conclusion}*
Unordnung *f* chaos, disorder
unscharf fuzzy
  unscharfe Information fuzzy information
  unscharfe Logik *f* fuzzy logic
  unscharfe Menge *f* fuzzy set
  unscharfe Verteilung *f* fuzzy distribution
  unscharfer Algorithmus *m* fuzzy algorithm
  unscharfer Automat *m* fuzzy automaton
  unscharfer Quantor *m* fuzzy quantor
  unscharfes System *n* fuzzy system
Unschärfe *f* fuzziness
Unsicherheit *f* uncertainty, error
Unsicherheitsintervall *n* uncertainty interval
unsichtbar machen hide/to
unstabil astable
unstetig unsteady
Unstetigkeit *f* unsteadiness
unsymmetrisch non-symmetric[al]
unteraktiv hypoactive

Unteraufgabe *f* subtask
unterbrechbar interruptable
Unterbrechbarkeit *f* interruptability
unterbrechen break/to; interrupt/to *{program}*
Unterbrechung *f* interruption *{program}*
Unterbrechungsstelle *f* breakpoint
untergeordnet secondary, subordinated
untergeordnetes System *n* slave (subordinated) system, subsystem
untergeordnetes Ziel *n* subgoal
Untergruppe *f* subgroup
Unterklasse *f* subclass
Unterkörper *m* subfield
Untermatrix *f* submatrix
Untermenge *f* subset
Untermodell *n* submodel
Unternehmensforschung *f* operations research
Unternehmensspiel *n* management game
unterscheidbares Ereignis *n* discernible event
unterscheiden distinguish/to, discern/to
Unterscheidung *f* **in der Verhaltensweise** discrimination *{if two stimuli are present and one of them is reinforced the latter does provoke a response and the non-reinforced one does not}*
Unterschriftsbestätigung *f* signature verification
Untersystem *n* subsystem
unterteiltes System *n* subdivided system
Unterteilung *f* subdivision
Untervektorraum *m* subvector space

Unterweisung *f* instruction
Unterziel *n* subgoal
unüberwachtes Lernen *n* unsupervised learning
ununterbrochen continuous
unvereinbar inconsistent
unvereinbare Ereignisse *npl* inconsistent (incompatible) events
unverzweigt non-branched
unverzweigter Algorithmus *m* non-branched algorithm
unvollständig incomplete; fuzzy
unvollständige Aussage *f* incomplete statement
unvollständige Induktion *f* incomplete induction
unvollständige Information *f* incomplete information
Unvollständigkeitssatz *m* incompleteness theorem
unvorhersehbar non-predictive
unwahr false *{logic}*
unwesentliches Spiel *n* unessential game
unwesentliches Wort *n* [in einem Kontext] stopword, noiseword
unzerlegbar indecomposable *{logic expression}*
unzulässig inadmissible, illegal, invalid, *{s.a. verboten}*
unzulässige Folge *f* illegal sequence *{e.g. switching sequence}*
unzulässige Lösung *f* non-feasible solution
unzulässige Strategie *f* non-feasible strategy
unzulässiger Ablauf *m* illegal sequence
unzulässiges Zeichen *n* illegal character
Ursache *f* cause

Ursprung *m* origin
Ursprungssprache *f* source language

V

vage fuzzy
vages System *n* fuzzy system
Valenz *f* valence
Valenz *f* **des Knotens** valence of node *{graph}*
Validierungsparameter *m* validation parameter
Validität *f* validity *{prognosis}*
Variable *f* variable
Variable *f* **der verflossenen Zeit** age variable
Variablenverschmelzung *f* merging of variables
Variation *f* **der Parameter** parameter variation
Vektor *m* vector *{ordered list of numbers}*
verallgemeinern generalize/to
verallgemeinert generalized
verallgemeinerte Koordinaten *fpl* generalized coordinates
verallgemeinerter Algorithmus *m* general algorithm
verallgemeinerter Graph *m* generalized graph
Verallgemeinerung *f* generalization, generality
Veränderbarkeit *n* **des Maßstabes** scalability, scalability
Veränderung *f* change
Veränderung *f* **infolge technischer Weiterentwicklung** migration
Verarbeitung *f* **natürlicher Sprache** natural-language processing

Verarbeitung *f* **von Gedanken** thought processing
Verarbeitungsabschnitt *m* processing section
Verarbeitungselement *n* processing element *{ generally a neuron, it has a number of inputs and a single output}*
Verarbeitungsknoten *m* processing node
verbergen conceal/to; hide/to
verbinden interlink/to
Verbinder *m* link
Verbindung *f* combination; joint; interconnection *{the path from one neuron to another one}*; interlinking *{of systems}*
Verbindung lösen unlink/to
Verbindungsaufbau *m* combination
Verbindungselement *n* link
Verbindungsstück *n* joint
verblocken interlock/to
verborgen hidden
  verborgene Aufbewahrung *f* hiding
  verborgene Schicht *f* hidden layer
  verborgenes Neuron *n* *s.* verdecktes Neuron
verbotene Kombination *f* forbidden combination
verbotener Zustand *m* forbidden state
Verbundwahrscheinlichkeit *f* joint probability
verdeckte Linie *f* hidden line
  verdecktes Neuron *n* hidden neuron *{one of all the neurons between the input ones and the output ones}*
Verdichtungspunkt *m* accumulation point *{topology}*
Vereinbarung *f* convention, declarative sentence
vereinfachen simplify/to *{e.g. network}*
vereinigen unite/to, merge/to
Vereinigungsmengenaxiom *n* union axiom
Vererbung *f* inheritance
Vererbungshierarchie *f* inheritance hierarchy
Verfahren *n* **der verallgemeinerten Gradienten** method of generalized gradients
Verfahrensdarstellung *f* procedural representation
Verfahrensforschung *f* operations research
Verfahrensmodell *n* procedural model
verfahrensorientierte Sprache procedure-oriented language
verfahrensorientiertes Programmiersystem *n* procedure-oriented programming system
verfahrensorientiertes Programmpaket *n* procedure-oriented software package
verflochtene Netzpläne meshed graphs
verfolgen pursue/to
Verfolger *m* pursuer
Verfolgter *m* evader
Verfolgungsbedingung *f* evasion condition
Verfolgungsspiel *n* evasion game, pursuit game
Verfolgungsstrategie *f* evasion strategy
verfügbar available
verfügbare Zeit *f* available time
Verfügbarkeit *f* availability
Vergleich *m* comparison

vergleichen compare/to, match/to
Vergleichsabfrage *f* relational query
Vergleichsmodell *n* reference model
Vergleichsoperator *m* relational operator
Verhalten *n* behaviour; response *{after a stimulus}*
Verhalten *n* **von Automaten** behaviour of automata
Verhaltensanalyse *f* behaviour analysis
Verhaltensmodell *n* behaviour model
Verhaltensmuster *n* pattern of behaviour
Verhaltensschwelle *f* behavioural threshold
Verhaltenssystem *n* behaving system
Verhaltensweise *f* behaviour
Verhältnis *n* proportion
verhindern prevent/to, inhibit/to
Verkehr *m* traffic *{queueing theory}*
Verkehrsanalyse *f* analysis of traffic flow, traffic analysis
Verkehrsleitcomputer *m* traffic control computer
Verkehrsstauung *f* congestion
Verkehrssteuercomputer *m* traffic control computer
Verkehrswert *m* traffic value *{queueing theory}*
verketten chain/to, interlink/to, link/to; concatenate/to
verkettet linked
   verkettete Datenmenge *f* concatenated data set
Verkettung *f* interlinking
   Verkettung *f* **von Aussagen** chaining
   Verkettung *f* **von Daten** concatenation
   Verkettung *f* **von Wissensbasen** knowledge-base chaining
Verkettungsoperation *f* linkage operation
verknüpfen link/to *{e.g. graphs}*
verknüpfen/über ODER-Gatter OR/to
Verknüpfung connection, link; combination
Verknüpfungsbefehl *m* connective instruction
Verknüpfungselement *n* logic element
Verknüpfungsglied *n* switching (combinational) element
Verknüpfungsgröße *f* connecting quantity
Verknüpfungslinie *f* connecting line
Verknüpfungsoperation *f* connective operation *{switching algebra}*
Verknüpfungspartikel *n* connective particle
Verknüpfungsschaltung *f* combinational circuit, combinatorial network
Verknüpfungsstelle *f* connection point
Verlängerung *f* extension
Verlauf *m* behaviour *{e.g. curve}*
Verletzung *f* **von Randbedingungen** constraint violation
Verlustfunktion *f* loss function
Verlustprozeß *m* loss process
Verlustsystem *n* loss system
Verlustverkehr *m* loss traffic
Verlustwahrscheinlichkeit *f* loss probability
vermascht complex, meshed
   vermaschte Strukturen *fpl* interconnected (meshed) structures
Vermaschungsgrad *m* complexity

vernetzt *s.* vermascht
verrauscht noisy
verrauschtes Signal *n* noise-embedded (noisy) signal
verrauschtes System *n* noisy system
verrichten act/to, perform/to
verriegeln interlock/to
verschachtelte Rekursion *f* interlocked recursion
verschiebbar relocatable
verschiebbarer Ausdruck *m* relocatable expression
Verschiebbarkeit *f* relocatability
verschieben move/to
verschieblicher Term *m* relocatable term
Verschieblichkeit *f* relocatability
Verschiebung *f* shift
Verschiebung *f* **nach links** shift left
Verschiebung *f* **nach rechts** shift right
verschlüsselter Text *m* cyphertext
Verschmelzungsgraph *m* merger graph
Verschmelzungsregel *f* contraction rule
Verschmelzungstabelle *f* merger table
versinnbildlichen symbolize/to
Verständlichkeit *f* intelligibility
verstärken amplify/to; strengthen/to *{stimulus}*
verstärkender Reiz *m* reinforcer *{a stimulus, as a reward, that strengthens a desired response}*
Verstärkung *f* gain; reinforcement
Verstärkung *f* **des Adaptivkreises** adaptive-loop gain
verstecken conceal/to

verstecktes Element *n* hidden element
verstellen control/to
Verstopfung *f* congestion
Versuch *m* trial, try
Versuch-und-Irrtum-Methode *f* trial-and-error method
versuchen try/to
Versuchsergebnis *n* experimental result
Versuchsmodell *n* developmental model
Versuchsplan *m* experimental plan
Versuchsplanung *f* experiment designing (planning)
verteilt distributed
verteilte Intelligenz *f* distributed intelligence
verteilte Logik *f* distributed logic
verteilte Steuerung *f* distributed control
verteiltes Gedächtnis *n* distributed memory *{the memory of a neural network is not located at a single address, but is spread throughout the entire parallel system}*
verteiltes System *n* distributed system
Verteilungsdichte *f* probability density
Verteilungsdichtefunktion *f* density function
verteilungsfreier Test *m* non-parametric test
Verteilungsgesetz *n* distribution law
Verteilungsproblem *n* distribution problem, transport[ation] problem
verträglich compatibel; consistent
Verträglichkeit *f* compatibility
Vervielfältiger *m* duplicator

Verwaltung *f* management
Verwaltung *f* **der Wissensbasis** knowledge base management
Verwaltungssystem *n* management system
Verweilprozeß *m* holding process
Verweilzeit *f* detention time
Verwendbarkeit *f* usability, adaptability
verzerrungsfrei linear
Verzögerung *f* delay
Verzögerung *f* **von Entscheidungen zur Problemlösung [bis möglichst viele Informationen vorliegen]** least commitment
verzweigen branch/to
verzweigter Prozeß *m* tree-like process, branched process
Verzweigung *f* branch[ing]; ramification (graph)
Verzweigungsalgorithmus *m* branch-and-bound algorithm
Verzweigungsmethode *f* branch-and-bound method, branch method
Verzweigungsprinzip *n* branch-and-bound principle, branching principle
Verzweigungspunkt *m* branch[ing] point, ramification point
Verzweigungsstelle *f* branch[ing] point
vieldeutig ambiguous
Vieldeutigkeit *f* ambiguity
Vielfachleitung *f* trunk
Vierfachgatter *n* **mit je zwei Eingängen** two-input quad gate
Vierfarbenproblem *n* four-colour problem
Vierpersonenspiel *n* four-person game
vierwertige Logik *f* four-valued logic, quaternary logic

virtuelles Bild *n* virtual image {projected image}
Vollalternative *f* elementary alternative, maxterm
vollbeschränkter Raum *m* precompact space
vollbeschränktes Spiel *n* precompact game
voll full
Vollkonjunktion *f* elementary conjunction, minterm
Vollordnungsrelation *f* total ordering relation
vollständig complete, total
Vollständigkeit *f* completeness
Volltexterfassung *f* full acquisition
von-Neumann-Computer *m* conventional computer
von-Neumann-Verarbeitung *f* sequential processing
vorangehende Aktivität *f* preceding activity {graph}
vorangepaßtes System *n* prematched system
voranschreitende Reduktion *f* progressive reduction {logic}
vorausberechnen precompute/to, predetermine/to, predict/to
vorausberechneter Wert *m* predicted value
Vorausbestimmung *f* predetermination
Voraussage *f* prediction
voraussagen predict/to
Voraussetzung *f* premise, assumption {logic}
Vorbedingung *f* precondition
vorbereiten condition/to
vorbestimmen predetermine/to
vorbestimmt predetermined
Vorderglied *n* antecedent {logic}

Vordersatz *m* premise *{logic}*
Voreilung *f* advance *{phase}*
Voreilwinkel *m* advance angle *{phase}*
Vorentscheidung *f* predecision
Vorentwicklung *f* advance development
Vorereignis *n* pre-event *{network technique}*
Vorgabewert *m* default value
Vorgang *m* action, event, occurence, phenomenon, process; activity *{network technique}*
Vorgänger *m* parent *{search tree}*; antecedent *{logic}*
Vorgänger-Nachfolger-Beziehung *f* parent-child relation
Vorgängerknoten *m* parent node
Vorgängerzustand *m* predecessor state
Vorgangsknoten *m* occurrence node
Vorgangsknotennetz *n* activity-on-node network, activity-oriented network
Vorgangsnetz *n* activity network
Vorgangspfeil *m* occurrence arrow
Vorgangspfeilnetz *n* activity-on-arrow network, event-oriented network
vorgangspfeilorientierter Netzplan *m* activity-on-arrow network
Vorgangszeit *f* action time
vorgeben predetermine/to *{e.g. set value}*
vorgegeben definite
vorgegebene Gesetzmäßigkeit *f* desired law
vorgegebene Zeitfolge *f* timing
vorgegebener Wert *m* default value; set-point value
vorgegebener Zustand *m* predetermined state
Vorgeschichte *f* antecedent
Vorhalt *m* prediction
Vorhaltfilter *n* predicting filter
Vorhandensein *n* presence
vorhergesagter Wert *m* predicted value
vorhersagbar predictable
Vorhersage *f* forecast[ing], prediction
Vorhersagealgorithmus *m* prediction algorithm
Vorhersageeinrichtung *f* predictor
Vorhersagemethode *f* prediction method
Vorhersagemodell *n* predictive model
vorhersagen predict/to
vorhersagende Codierung *f* predictive coding
vorhersagendes Filter *n* predicting filter
Vorhersageproblem *n* prediction problem
Vorhersagestrategie *f* predictive strategy
Vorhersagesystem *n* prognostic system
Vorhersagetheorie *f* prediction theory of random processes
Vorhersagewert *m* predictive value
vorhersehbar predictive
Vorknoten *m* prenode
vorkommen occur/to
Vorlage *f* model
Voroptimierung *f* pre-optimization
Vorrang *m* priority
Vorrangebene *f* priority level
Vorrangentscheidung *f* priority decision
Vorschub *m* advance

Vorsichtsstrategie *f* minimax strategy
vorwärts forward
Vorwärtskopplung *f* feedforward *{neurons take their inputs only from the previous layer and send their outputs only to the next layer}*
vorwärtsverkettetes Shell *n* inductive shell
Vorwärtsverkettung *f* forward chaining
Vorwärtszweig *m* forward branch
Vorwiderstand *m* additional resistance
Vorzugsrelation *f* preference

W

wachsen/über alle Grenzen grow without bound/to
wachsender Automat *m* growing automaton
Wachstum *n* growth
Wachstumskurve *f* growth curve
Wachstumsmodell *n* growth model
wahlfrei random *{e.g.access}*
wahllos verteilen randomize/to
wahr true, real; tautological *{propositional logic}*
wahre Aussage *f* true [declarative] sentence
wahre Schlußfolgerung *f* true conclusion
Wahrheitsbegriff *m* truth concept
Wahrheitserhaltung *f* truth maintenance
Wahrheitskonzept *n* truth concept
Wahrheitsschwelle *f* truth threshold
Wahrheitstabelle *f* truth table
Wahrheitstafel *f* truth table

Wahrheitsvorwärtsschluß *m* modus ponens *{lat. artificial word}*
Wahrheitswert *m* truth value
Wahrheitswerteanalyse *f* truth-value analysis
Wahrheitswertefunktion *f* truth function
Wahrheitswertelogik *f* propositional logic
Wahrheitswertetabelle *f* truth table
Wahrheitswertetafel *f* truth table, Boolean operation table
Wahrheitswertetafelgenerator *m* truth table generator
Wahrheitswertevariable *f* truth variable
wahrnehmbar perceptible
Wahrnehmbarkeit *f* perceptibility
wahrnehmen percept/to
Wahrnehmung *f* perception
Wahrnehmungselement *n* perceptor
Wahrscheinlichkeit *f* probability
Wahrscheinlichkeit *f* **des Auftretens** occurrence probability
Wahrscheinlichkeit *f* **des sicheren Auftretens [eines Ereignisses]** probability of certainty
Wahrscheinlichkeitsautomat *m* probabilistic automaton
Wahrscheinlichkeitsbeziehung *f* probabilistic relationship
Wahrscheinlichkeitsdichte *f* probability density
Wahrscheinlichkeitsfeld *n* probability array
Wahrscheinlichkeitskurve *f* probability curve
Wahrscheinlichkeitslogik *f* probabilistic logic
Wahrscheinlichkeitsmaschine *f* probabilistic machine

Wahrscheinlichkeitsmodell *n* probability model
Wahrscheinlichkeitsprozeß *m* probabilistic process
Wahrscheinlichkeitsraum *m* probability space
Wahrscheinlichkeitsrechnung *f* calculus of probability
Wahrscheinlichkeitsschwelle *f* probability threshold
wahrscheinlichkeitstheoretisch probabilistic {s.a. probabilistisch}
wahrscheinlichkeitstheoretische Verkehrsanalyse *f* probabilistic traffic analysis
wahrscheinlichkeitstheoretisches Verkehrsmodell *n* probabilistic traffic model
Wahrscheinlichkeitstheorie *f* probability theory
Wahrscheinlichkeitsvergleich *m* probability comparison
Wald *m* forest {graph}
warten pause/to
Warteschlange *f* chain, queue
  Warteschlange bilden queue/to
Warteschlangenlänge *f* queue length
Warteschlangenproblem *n* queueing problem
Warteschlangenprogramm *n* **für die Bearbeitung einer Liste von beiden Enden her** double-ended queue [program] {allows e.g. insertions at both ends of a list}
Warteschlangensystem *n* queueing system
Warteschlangensystem *n* **mit [nur] einer Leitung** single line queue system
Warteschlangentheorie *f* congestion theory, queueing theory

Wartesystem *n* waiting system
Wartesystem *n* **mit Beschränkungen** constrained waiting system
Wartesystem *n* **mit [nur] einer Leitung** *f* single-line waiting system
Wartewahrscheinlichkeit *f* probability of waiting
Wartung *f* updating; maintenance
Wechsel *m* change
Wechselinformation *f* transinformation
wechselseitige Abhängigkeit *f* interdependence
wechselseitiger Ausschluß *m* mutual exclusion
wechselseitiger Zusammenhang *m* interdependence
Wechselwirkung *f* interaction
wechselwirkungsfrei non-interactive
WEDER-NOCH NEITHER-NOR
Weg *m* approach; route; path {track through a state graph}
Wegeermittlung *f* routing
Weginformation *f* geometrical information {numerical control}
Wegsteuerung *f* path control {robot}
Wegverfolger *m* router
Wegverfolgungscode *m* router code
weiche Beschränkung soft restriction
  weiche Restriktion *f* soft restriction
Weise *f* mode
Weiterbewegung *f* advance
weiterentwickeltes Betriebssystem *n* advanced operating system
weiterentwickeltes Logikverarbeitungssystem *n* advanced logic processing system
weitestmögliche Entscheidungsverschiebung *f* least commitment

Weitschweifigkeit *f* redundancy
Welt *f* world
Weltmodell *n* world model
Weltwissen *n* world knowledge
WENN (ONLY THEN) IF
WENN DANN IF-THEN
WENN-DANN-verknüpfen imply/to
WENN-DANN-Verknüpfung *f* implication
WENN SO seq
WENN-UND-NUR-WENN-Operation *f* equivalence operation
Werkstückhandhabungsautomat *m* parts-handling machine *{robot}*
Werkzeug *n* tool
Wert *m* amount; value *{of a variable}*; magnitude *{length of a vector given by the square root of the sum obtained by adding the squares of the vector components}*
Wert *m* **der Zustandsgröße** state value
Wert *m* **des Spiels** game value
Wertebereich *m* data range
Wertereihe *f* range
Wertigkeit *f* priority level; valence *{e.g. graph}*
Wettfahrt *f* race
Wettlauf *m* racing, race, hazard, hunting
Wettlaufbedingungen *fpl* race conditions
widerlegen disprove/to
Widerlegung *f* disproving *{hypothesis}*
Widerlegungsregel *f* modus tollens *{lat., artificial expression in deductive logic}*
Widerspruch *m* contradiction *{logic}*

Widerspruch *m* **mit sich selbst** antinomy *{logic}*
widersprüchlich contradictory
widerspruchsfrei non-contradictory
Widerspruchsfreiheit *f* consistency
Widerspruchsfreiheit *f* **von Axiomensystemen** consistency of axiom systems
wiederauffinden retrieve/to
Wiederauffinden *n* **von Informationen** information retrieval
wiederfinden *s.* wiederauffinden
wiedergewinnen retrieve/to
Wiederholbarkeit *f* repeatability
wiederholen repeat/to; playback/to
Wiederholfunktion *f* cycling
wiederkehrend recurrent
willkürlich random; arbitrary
willkürliche Größe *f* arbitrary quantity
willkürliches Signal *n* arbitrary signal
Windung *f* turn
winkeltreu conformal
wirken auf actuate/to
wirklich actual, real
wirkliches Problem *n* real-world problem
wirksam effective, active
Wirksamkeit *f* **eines Systems** system effectiveness
Wirkschema *n* actual operating diagram
Wirkung *f* effect, action
Wirkungsbereich *m* domain of influence
Wirkungselement *n* actor
Wirkungsgrad *m* efficiency
Wirkungsgröße *f* actuating quantity
Wirkungskette *f* action chain
Wirkungsweg *m* actuating path

Wirkungsweise f action
Wissen n knowledge; expertise
Wissen n des Menschen human knowledge
Wissen n über das Wissen metaknowledge
Wissen n über Wissen in anderen Ebenen meta-level knowledge
Wissensakquisition f knowledge acquisition
wissensbasierte Steuerung f knowledge-based control
wissensbasiertes System n knowledge-based system
Wissensbasis f knowledge base
wissenschaftlich-technische Prognostizierung f von Produktionsmodellen scientific and technical production model prognostication
wissenschaftliche Information f scientific information
Wissensdarstellung f knowledge representation
Wissensdarstellung f mittels logischer Verknüpfungen logic representation
Wissenserfassung f knowledge acquisition
Wissenserwerb m knowledge acquisition
Wissensgebiet n knowledge domain
Wissenskompilierung f knowledge proceduralization (operationalization)
Wissensquelle f knowledge source
Wissensrelevanz f knowledge relevance
Wissenssenke f knowledge sink
Wissensstand m knowledge level
Wissenstechnik f knowledge engineering

Wissensverarbeitung f knowledge processing
Wissensverarbeitungssystem n knowledge information processing system
Wissensverfügbarkeit f knowledge availability
Wörterbuch n dictionary, glossary
Wörterbuchsuche f dictionary retrieval
Wurzel f root
Wurzelknoten m root node {tree structure}

Z

Zahlentheorie f number theory
Zahlenwert m figure
Zähler m counter, accumulator
Zahlkörper m number field
zehnwertige Logik f denary logic
Zeichen n character, figure, symbol
Zeichenautomat m automatic drafting machine
Zeichenerkennbarkeit f intelligibility
Zeichenerkennung f character recognition; pattern recognition
Zeichenreihe f sequence of symbols
Zeichensatz m character set
Zeichenvorrat m alphabet
Zeitablauf m timing
Zeitablaufplanung f time scheduling
Zeitdauer f eines Vorgangs action time
Zeitdehnung f time scaling
zeitdiskretes Modell n discrete-time model
Zeitfolge f timing

Zeitkoordinate *f* time coordinate
zeitlich konstant constant with time
zeitliche Koordinate *f* time coordinate
zeitliches Verhalten *n* time behaviour
zeitoptimales System *n* optimum-time system
zeitparallel serial
Zeitplan *m* schedule
Zeitraffung *f* time scaling
Zeitscheibe *f* time slice
Zeitsteuerung *f* timing
Zeitteilbetrieb *m* time sharing
zeittreue Nachführung *f* updating
zeittreue Simulation *f* event-by-event simulation
zeitunabhängig time-independent
Zeitverhalten *n* time behaviour, timing
Zeitverlauf *m* time behaviour
Zeitvorrat *m* available time
zentral central
zentralisiert centralized
zentrieren centre/to
zerlegbar decomposable *{e.g. expression}*
Zerlegbarkeit *f* decomposability
Zerlegbarkeitsbedingung *f* decomposability condition
Zerlegung *f* decomposition
Ziel *n* aim, destination, objective, goal, target
Zielbaummethode *f* relevance tree method
Zielbedingung *f* objective condition
Zieldaten *npl* destination data
Zielebene *f* target level
Zielereignis *n* objective event
Zielfunktion *f* target function
Zielfunktionsvektor *m* objective function vector
zielgerichtetes Schließen *n* goal-directed inference
zielgesteuerte Inferenz *f* goal-directed inference
Zielgröße *f* objective quantity
Zielhierarchie *f* objective hierarchy
Zielkoordination *f* objective coordination
Zielmenge *f* objective set
Zielniveau *n* target level
Zielort *m* destination
Zielphase *f* target phase
Zielregression *f* goal regression *{using of subgoals}*
Zielsatz *m* target set
Zielsprache *f* target language
Zielsteuerung *f* goal driving *{problem solving}*
zielsuchendes System *n* goal-searching system
Zielwert *m* objective value
Zielzustand *m* goal state
Ziffer *f* figure
zufällig accidental, random, stochastic; *{s.a. stochastisch}*
zufällige Auswahl *f* random choice
zufällige Funktion *f* random function
zufällige Größe *f* random quantity
zufällige Suche *f* random search
zufällige Variable random variable
zufälliges Ereignis *n* accident, random event
Zufälligkeit *f* randomness
Zufall *m* accident, chance
zufallsabhängig accidental
zufallsabhängiges Signal *n* accidental signal
Zufallsbewegung *f* random walk
Zufallscodierung *f* random coding

Zufallsereignis *n*   random event
Zufallsfunktion *f*   random function
Zufallsgesetz *n*   law of chance
Zufallsgröße *f*   random quantity, random variable
Zufallslogik *f*   random logic
Zufallsmethode *f*   hazard method
Zufallsprozeß *m*   random process
Zufallssuche *f*   random search [method]
Zufallsvariable *f*   random variable
Zufallsvektor *m*   random vector
Zufallswegproblem *n*   random-path problem
Zugriffscodierung *f*   access coding
Zugriffsmechanismus   access mechanism
Zugriffszeit *f*   access time
Zugriffsziel *n*   access destination
Zugriffszustand *m*   access state
zulässig   acceptable, admissible, allowed, allowable, feasible, legitimate, valid
zulässig färbbarer Graph *m*   valid colourable graph
zulässige Lösung *f*   admissible solution
zulässige Menge *f*   admissible set
zulässige Strategie *f*   feasible strategy, permissible strategy, valid strategy
zulässige Trajektorie *f*   admissible trajectory
zulässiger Bereich *m*   feasible region *{e.g. of a function}*
zulässiger Punkt *m*   admissible point
zulässiger Vektor *m*   admissible vector
zulässiger Weg *m*   admissible path *{graph technique}*

zulässiger Wert *m*   admissible value
zulässiges Gebiet *n*   admissible set
Zulässigkeit *f*   admissibility, feasibility
zünden   fire/to *{e.g. neuron}*
zunehmender Stellenwert *m*   ascending order of significance
zuordnen   allocate/to, assign/to; associate/to *{values}*; match/to
Zuordner *m*   translator, interpreter
Zuordnung *f*   assignment, correspondence; dedication
Zuordnungsproblem *n*   assignment problem, coordination problem
zurückholen   retrieve/to
Zurückverfolgen *n*   backtracking
zusammenballen   chunk/to *{build-up of collections of stimuli and responses into behaviour-controlling units}*
Zusammenfassung *f*   abstract
zusammengesetzt   complex
zusammenhängend   path-connected
zusammenhängender Automat *m*   connected automaton
zusammenhängender Graph *m*   connected graph, continuous graph
Zusammenhang *m*   connectivity
Zusammenhang *m* **eines Graphen**   connectedness of a graph
zusammenpassen   match/to
zusammenschrumpfen   shrink/to
zusammentreffend   concurrent
Zusatzlogik *f*   additional logic
Zuschnittproblem *n*   cutting (tailoring, trimming) problem
Zusicherung *f*   assertion
Zustand *m*   state, condition
zustandsabhängig   state-dependent
Zustandsänderung *f*   change in state

Zustandsbahn *f* [state] trajectory
Zustandsbeobachter *m* state observer
Zustandsbeschränkung *f* constraint of state variables
Zustandsebene *f* state plane
Zustandserkennung *f* state recognition
Zustandsgleichung *f* state [variable] equation
Zustandsgraph *m* state graph
Zustandsgröße *f* parameter, state variable
Zustandsgrößengleichung *f* state variable equation
Zustandsgrößenvektor *m* state variable vector
Zustandskurve *f* state curve
zustandslinear state-linear
Zustandsmenge *f* state set
Zustandsraum *m* state space
Zustandsraumanalyse *f* state space analysis
Zustandsraummethode *f* state space method
Zustandsraumtrajektorie *f* state space trajectory
Zustandsreduktion *f* state reduction
Zustandsschätzer *m* in der höheren **Ebene** *f* higher-level estimator
Zustandsschätzer *m* in der unteren **Ebene** *f* lower-level estimator
Zustandsschätzung *f* state estimation
Zustandstabelle *f* state table
Zustandstrajektorie *f* state trajectory
Zustandsübergangsmatrix *f* state transition matrix
Zustandsüberwachung *f* condition monitoring

zustandsunabhängig state-independent
zustandsunabhängige Partition *f* state-independent partition
Zustandsvariable *f* state variable
Zustandsvektor *m* status vector
Zustandsveränderliche *f* state variable
Zustandswahrscheinlichkeit *f* state probability
zuteilen allocate/to
Zuteilungsproblem *n* apportioning problem, assignment problem *{optimization}*
Zuverlässigkeit *f* reliability
Zuverlässigkeit *f* **des Systems** system reliability
Zuverlässigkeitsberechnung *f* reliability calculation
zuweisen allocate/to, assign/to
Zuweisung *f* assignment
Zwangsbedingung *f* constraint
Zwangsbedingungshyperfläche *f* constraint hypersurface
zweideutig ambiguous
Zweideutigkeit *f* ambiguity
Zweiebenensystem *n* two-level system
Zweig *m* branch; path *{network}*
Zweipersonen-Nichtnullsummenspiel *n* two-person non-zero sum game
Zweipersonen-Nullsummenspiel *n* two-person zero-sum game, matrix game
Zweipersonenspiel *n* two-person game
Zweiphasenbedienungssystem *n* two-phase queueing system
Zweischleifenadaption *f* two-loop adapt[at]ion

zweischleifige Adaption f two-loop adapt[at]ion
Zweisortentransportproblem n two-sort transportation problem
zweistellig binary
zweiwertig binary, dual; two-valued
  zweiwertige Funktion f two-valued function
  zweiwertige Logik f binary logic, two-valued logic
Zwischenankunftszeit f [inter-] arrival time *{queueing theory}*
Zwischenkörper m intermediate field *{mathematics}*
Zwischensprache f intermediate language
Zwischenverbindungssystem n link system
Zwischenziel n subgoal
Zwischenzustand m intermediate state
zyklisch cyclic
  zyklische Folge f cyclic sequence
  zyklische Gruppe f cyclic group *{group theory}*
  zyklische Verschiebung f cyclic shift
Zyklus m cycle *{e.g. graph or program}*

Junge, H.-D.

parat Wörterbuch Informationstechnologie
Deutsch/Englisch

parat Dictionary of Information Technology
German/English

1990. VIII, 694 Seiten. Gebunden. DM 195,-. ISBN 3-527-26420-5 (VCH)

parat Dictionary of Information Technology
English/German

parat Wörterbuch Informationstechnologie
Englisch/Deutsch

1989. VIII, 927 Seiten. Gebunden. DM 195,-. ISBN 3-527-26430-2 (VCH)

Jedes dieser beiden Wörterbücher enthält mehr als 50 000 Einträge, dazu zahlreiche ergänzende Hinweise auf Synonyme, Fachgebiete, Erläuterungen, Phrasen und Anwendungsbeispiele, ohne die der Übersetzer gerade auf diesem Gebiet nicht auskommt.

Behandelt werden praktisch alle Bauelemente, Geräte und Verfahren der High-Technology und Automation, insbesondere Sensorik, Wandlung, Codierung sowie Speicherung und Verarbeitung von Daten bis hin zur Robotik und künstlichen Intelligenz, ferner Gebiete wie Zuverlässigkeitstechnik und Statistik, kurz alles, was eine moderne technische Übersetzung erfordert. Quellen sind neben den nationalen und internationalen Standards die modernste Fachliteratur sowie neueste Firmenschriften, Gerätebeschreibungen, Bedienungsanleitungen und vieles andere mehr.

Preisänderung vorbehalten
Stand der Daten: November 1990

Schenk, H.

parat Dictionnaire de la Robotique
Français/Allemand

parat Wörterbuch Robotik
Deutsch/Französisch

1989. X, 214 Seiten. Gebunden. DM 68,-. ISBN 3-527-26940-1 (VCH)

Mehr als 3000 Einträge in jeder Sprachrichtung geben einen Überblick über den Wortschatz für das neue Gebiet der Robotertechnik. Sie sind in vielen Fällen ergänzt um kurze Erläuterungen, zusätzliche Angaben zu Herkunft und Umfeld, Anwendungsbeispiele und Verweise auf Synonyme.

Junge, H.D.

parat Dictionary of Measurement Engineering and Units
English/German

parat Wörterbuch Meßtechnik und Einheiten
Deutsch/Englisch

1990. Ca. XII, 594 Seiten. Gebunden. Ca. DM 124,-.
ISBN 3-527-26422-1 (VCH)

Mit etwa 10000 Einträgen je Sprachrichtung ist dieses Wörterbuch ein wichtiges Hilfsmittel für jeden technischen Übersetzer. Denn gerade für die immer aktueller werdende und immer weiter verbreitete Automatisierungstechnik bildet die Meßtechnik - von Sensorik bis zur Meßdatenverarbeitung - die Grundlage.
Darüber hinaus bietet dieses Werk eine umfangreiche Liste aller englischen, deutschen und SI-Einheiten sowie ihrer Umrechnungsfaktoren. Es deckt damit einen Bedarf, der durch die immer noch häufige Verwendung SI-fremder Einheiten in der technischen und wissenschaftlichen Literatur entsteht.

Preisänderung vorbehalten
Stand der Daten: November 1990

parat Index
of Acronyms and Abbreviations in
Electrical and Electronic Engineering

1989. V, 538 Seiten. Gebunden. DM 224,-. ISBN 3-527-26842-1 (VCH)

Dieser Index enthält etwa 45000 Einträge von Acronymen und Abkürzungen aus dem Bereich Elektrotechnik/Elektronik. Viele Abkürzungen sind genormt (IEC, IEEE und andere Organisationen), andere stammen aus dem Bereich privater Firmen und Institutionen oder aus dem militärischen Bereich.

Durzok, J.
parat Fachlexikon
Messung und Meßfehler

1989. VII, 172 Seiten mit 28 Abbildungen und 1 Tabelle.
Gebunden. DM 68,-. ISBN 3-527-26747-6 (VCH)

Dieses Lexikon enthält alle wichtigen Fachbegriffe der Lehre vom Messen mit Definitionen und Erläuterungen. Es wendet sich an Techniker, Ingenieure und Studenten aller technischen und naturwissenschaftlichen Fachrichtungen. Der Autor erklärt die Voraussetzungen des Messens, die Gemeinsamkeiten und Gesetzmäßigkeiten aller Meßmethoden und -geräte in leicht verständlicher, aber dennoch wissenschaftlich exakter Weise. Die Leser erhalten zahlreiche Informationen über die Aussagekraft von Ergebnissen unter den von ihnen gewählten Meßbedingungen. Mit dieser Darstellung vor allem der Grundlagen der Metrologie bleibt das Lexikon über einen längeren Zeitraum aktuell; dies ist auch deshalb gewährleistet, da der Autor sich nicht in Details über die verwirrende Anzahl von Meßmethoden verliert, mit denen Forschung und Industrie heute arbeiten.

Die Zuordnung angloamerikanischer Begriffe zu den deutschen Stichworten und die daraus abgeleitete englisch-deutsche Wortliste machen dieses Buch auch zu einem kleinen Wörterbuch der Metrologie.

Preisänderung vorbehalten
Stand der Daten: November 1990